Testing Large Ultra-Lightweight Spacecraft

Testing Large Ultra-Lightweight Spacecraft

EDITED BY

Jeremy A. Banik
Christopher H. Jenkins

Volume 253
Progress in Astronautics and Aeronautics

Timothy C. Lieuwen, Editor-in-Chief
Georgia Institute of Technology
Atlanta, Georgia

Published by
American Institute of Aeronautics and Astronautics, Inc.
12700 Sunrise Valley Drive, Suite 200, Reston, VA 20191-5807

Cover image: Courtesy NASA.

American Institute of Aeronautics and Astronautics, Inc., Reston, VA.

ISBN 978-1-62410-464-0

TABLE OF CONTENTS

PREFACE

The present *Testing Large Ultra-Lightweight Spacecraft* is the third volume in a series that began with *Gossamer Spacecraft: Membrane/Inflatable Structures Technology for Space Applications* (AIAA Progress in Astronautics and Aeronautics Series, vol. 191, C.H. Jenkins, Editor, 2001). That volume covered much of the then nascent technology, characterized by low Technology Readiness Levels (TRL), concept studies, and prototypes.

The second volume in the series, *Recent Advances in Gossamer Spacecraft* (AIAA Progress in Astronautics and Aeronautics Series, vol. 212, C.H. Jenkins, 2006), documented a maturing technology. Volume 212 presents detailed theoretical development, complex analysis, and the beginning of new experimental techniques.

In the intervening years since that second volume, the technology has continued to mature, as witnessed by:

- Deploying larger space platforms from smaller packed volumes, with fewer moving parts and more strain energy enforcements.
- Deployment articulations becoming more complex, with more degrees of freedom.
- Precision space structures growing larger, forcing flight qualification at the subscale level.

The present volume is focused on documenting the novel testing approaches that are unique to large ultra-lightweight space platforms. The scope of this book is to provide both the theory and especially the practice of testing large ultra-lightweight spacecraft for project managers and technical specialists. The intended outcome of the book is a single repository for all the unique approaches that we take in validating large ultra-lightweight space structures on the ground and in space. The expected audience includes engineers, technicians, and technology managers involved in test and measurement.

In addition to the primary structural categories, *large* and *ultra-lightweight*, another is evident within this volume. Namely, these structures are typically not self-supporting in the terrestrial gravity environment, requiring counterbalances or other measures to simulate zero-gravity in ground tests and during performance measurements.

This volume represents all types of large and ultra-lightweight architectures, mechanisms, and materials. For example, heritage metallic hinged architectures are represented alongside more advanced strain-energy composite deployed systems, and laminated plate apertures are included alongside tension-aligned membrane and mesh systems.

While a vast amount of information is included in this volume, there are several important topics not included due to time and availability. Notably among these are the requirements levied on facilities undertaking testing of large and ultra-lightweight space structures. Another is radiometric testing for thermal and optical performance. These topics will hopefully be covered in future editions of this volume.

The Editors wish to express their appreciation to the publications staff at AIAA for their help and guidance, which has been invaluable. We also owe a debt of gratitude to the SpaceCraft Structures Technical Committee of AIAA for their support of this project.

Finally, the Editors wish to thank the many expert authors who have contributed their vast knowledge and experience—and valuable time—to the production of the present volume. It has been our pleasure to work with them and learn from them. Without their expertise and dedication, this work would not have been possible.

Jeremy Banik
Chris Jenkins

CHAPTER 1

Introduction

Christopher H. Jenkins*

Montana State University

1.1 INTRODUCTION

This chapter first briefly reviews large ultra-lightweight spacecraft (LULS) and some relevant terminology. Next, it discusses activities known as experiment, test, and demonstration. Finally, it explores the Echo balloon satellites of the 1960s. These were the first LULS to undergo a significant test program.

1.2 LULS

Modern trends are driving spacecraft to the extremes of length scales. On the one hand, the ubiquitous trend toward miniaturization is pushing spacecraft to be ever smaller; on the other, many commercial, military, and science applications can be done only with very large systems. In either case, ultra-lightweight (or, *gossamer*) technology will enable a future generation of spacecraft that have remarkable capability to accomplish missions beyond the reach of current systems [1].

These ultra-lightweight spacecraft have many benefits and advantages over current mechanical systems. They are low in mass and can be packaged into small volumes, which can reduce the overall program cost by reducing the launch vehicle size. Reduction in total system mass and deployment complexity can also increase system reliability. Ultra-lightweight spacecraft are of particular advantage for mission architectures at the extreme ends of the size spectrum—from the very small to the very large. Examples of the former are inflatable solar arrays for small (micro-, pico-) satellites [1]. In the latter case,

---

*PhD, PE, Professor Emeritus

ultra-lightweight spacecraft technology will enable very large systems for bold missions of discovery such as

- Very large telescopes for imaging extra-solar planets, studying the formation of large-scale structure in the early universe, and continuously monitoring the Earth from distant vantage points.
- Large deployable and inflatable antennas for space-based radio astronomy, high-bandwidth communications from deep space, and Earth remote sensing with radar and radiometers.
- Solar sails for low-cost propulsion, station keeping in unstable orbits, and precursor interstellar exploration missions.
- Large solar power collection and transmission systems for human and robotic exploration missions, and for the commercial development of space.

We have before us the imprecise terms *large* and *ultra-lightweight*. For better or worse, there are no precise boundaries wherein LULS reside. It is possible, however, to provide some clarification regarding the scope of these spacecraft. For example, both of these terms may be placed in the context of space launch capacity.

Today, the maximum available launch capacity for payload diameter and mass is about 5 m and 3000 kg to geosynchronous orbit (GEO), respectively. So, going larger in space means going lighter and being more packageable. A figure of merit for "lightness" in some spacecraft is *areal density*. Taken simply, areal density is the aperture mass per unit of aperture surface area. Typically, ultra-lightweight refers to areal densities less than about 1 kg/m^2. (Non-aperture ultra-lightweight components, such as booms and trusses, necessarily use other metrics to define their lightness.)

A figure of merit for packaging is *packaging ratio*, the ratio of deployed to stowed volume. The stowed volume should not exceed that which would fit in the largest launch shroud. A packaging ratio of unity indicates the article did

TABLE 1.1 EXAMPLE PACKAGING RATIOS [2]

State of Practice (rank order for structural performance)	Deployed Size	Stowed Size	Packaging Ratio
JWST Primary	6.5 m	4.0 m	1.6:1
Exo-S Starshade	34 m	5.0 m	9:1
SkyTerra-1 Mesh Reflector	22 m	2.4 m	9:1
NG Telescopic Tube	33 m	2.4 m	14:1
ATK Graphite Coilable Boom	40 m	0.4 m	100:1
Graphite STEM	17 m	0.3 m	57:1

not need to be packaged for launch into space. By contrast, a packaging ratio of 100 shows that the deployed or service configuration volume is 100 times that of the stowed volume. For our purposes, "large" can be taken to mean a packaging ratio greater than unity. A sample of packaging ratios is given in Table 1.1. (Although a packaging ratio is formally a volume ratio, for practical purposes, packaging ratios may be couched in terms of ratios of characteristic lengths.)

1.3 EXPERIMENT, TEST, AND DEMONSTRATION

This book deals with a very physical part of space technology development that is variously referred to as *experiment*, *test*, or *demonstration*. Although there is considerable overlap in the use of these terms, they are far from the same thing.

Experiments are methods of pure research. Experiments are done to help understand what is not known, say, of some physical phenomenon. Of the three activities, the outcomes of experiments are the least deterministic: "In the case of an experiment, one may not even know what it means for it to work—the experiment has the capacity to surprise us" [3]. Given this high level of uncertainty, the experiment remains the most flexible procedurally. In the annals of science, experiment has led the way in developing theories that explain the world around us.

Like experiments, *tests* are also part of research, but they have more of an applied bent. Tests are formalistic procedures that produce data; the test formalism is often codified in standards. Thus, data from tests carry a higher degree of confidence than data from experiments, leading to increased results utilization [4].

A *demonstration* is "an event orchestrated to show how a process or product works [4]." "Demonstrations are designed to educate and convince once the exploration has been done and the discoveries have been made. Demonstrations have the power to convince because of the smoothness of performance, distancing the audience from the untidy craft of the scientist [3]."

Figure 1.1 places the three terms in the context of the maturity of technology development.

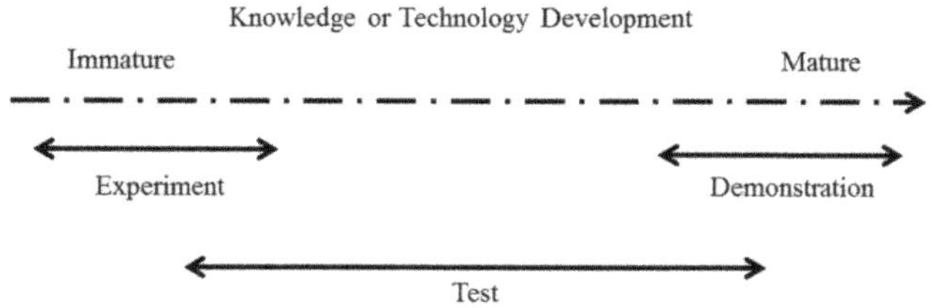

Fig. 1.1 The range of experiment, test, and demonstration relative to the maturity of knowledge or technology development.

Given this book's focus on practice, the term *test* will suffice to cover most of what needs to be discussed (see also Chapter 2).

1.4 IT ALL BEGAN WITH ECHO

The history of the Echo balloon satellites is inextricably linked to the complex history of satellites generally. According to Elder

> For some time, President Dwight D. Eisenhower had known that the intelligence-gathering agencies of the United States would soon have the ability to photograph the Earth with remarkable resolution from high altitude; his advisors also informed him that a satellite had excellent potential as a platform for basing such an observation system. Eisenhower understood, however, that under existing practice, any nation could consider a satellite passing over its territory as an invasion of its airspace and therefore could legally shoot the surveillance satellite down—if it possessed that capability. Unless the United States could somehow convince the nations of the world to regard satellites as having a different legal status than airplanes, the value of surveillance satellites would be severely limited.
>
> At this critical juncture, the world's scientific community had offered Eisenhower a possible way to establish the principle of legal satellite overflight. To enhance the gathering of useful data during the International Geophysical Year (IGY) 1957–58, these individuals in 1954 called on the governments of the world to launch satellites for conducting scientific experiments. American scientists did not hesitate to propose such a venture to the Eisenhower administration. Eager to provide a precedent for the overflight of other nations by surveillance satellites, Eisenhower approved the launching of an American scientific satellite in conjunction with the IGY. The official announcement came in July 1955.
>
> Soon after, the American coordinators of the IGY formed a committee to select experiments to include in the payload of a venture named Vanguard. One proposal presented to this group came from William J. O'Sullivan, Jr., of the National Advisory Committee for Aeronautics (NACA). He envisioned ejecting a small balloon out of the final stage of a launch vehicle, inflating it, and then observing the effects of atmospheric drag on the sphere. O'Sullivan suggested that this balloon could consist of either metallic foil or plastic with a metallic coating. The committee considered his proposal and shortly gave its approval [5].

O'Sullivan tells an interesting story about how he formulated the idea for the inflatable reflective sphere:

> In the year 1901, the Italian investigator Marconi sent a radio signal across the Atlantic Ocean. This was contrary to what was then believed to be the way in which radio signals traveled, and there was no explaining how this radio signal got around the curvature of the earth. This mystery continued for a full year until two researchers, Kennaly and Heaviside, realized that Marconi's signal must have been reflected by a layer of ionizing gas in the upper atmosphere.

> In the early days of radio, amateur operators competed to see how far they could "bounce" (reflect) radio waves off of the ionosphere. At sunrise or sunset, however, the reflective layers degrade and disappear, and communications by reflection cease. This breakdown led the radio amateurs to joke that "they needed a big mirror up in space to reflect back their signals when the ionosphere failed." This quip was apparently not lost on O'Sullivan [6].

Table 1.2 lists in chronological order the first nine satellites to be successfully launched and orbited. In terms of linear density alone (kg/m), Echo 1 began a new paradigm in LULS. In fact, if one computes the areal density of Echo 1 (mass/surface area) at about 22 g/m^2, this density rivals that of modern gossamer spacecraft such as the Inflatable Antenna Experiment ($\sim$14 g/m^2) [1].

Echo 1, America's first communications satellite, was managed by NASA's Jet Propulsion Laboratory. Figure 1.2 shows Echo 1 as it made its first pass over the Goldstone Tracking Station in Pasadena, California, in the early morning of August 12, 1960. Visible in this image are the movement of the antenna, star trails (shorter streaks), and Echo 1 (the long streak in the middle).

Echo 1 was made from a large number of Mylar gores that were 12 μm thick, coated with 2000 angstroms of vapor-deposited aluminum, and bonded together to form a sphere. The sphere was 30.5 m (100 ft) in diameter. For launch, the sphere was stowed in a spherical container that was 660 mm (26 in.) in diameter. To enable orbital deployment, 128 N (30 lb) of sublimating powders were sifted into the balloon structure. Four subscale inflation tests were made in the 12.5 m (41-ft.) diameter vacuum chamber at NASA Langley Research Center (LaRC). In addition, five free-space environmental tests were made in ballistic flights with rockets from NASA's Wallops Island Test Station. These tests were used to develop and validate mechanical packaging, ejection, and inflation techniques,

TABLE 1.2 SIZE AND MASS PROPERTIES OF THE FIRST NINE SATELLITES

Name	Launch Date	Size (m)	Mass (kg)	Mass/Size (kg/m)
Sputnik 1	4/10/1957	0.58	83.6	144
Sputnik 2	11/3/1957	4	508	127
Explorer 1	1/31/1958	2.06	13.97	6.78
Vanguard 1	3/17/1958	0.165	1.47	8.91
Explorer 3	3/26/1958	0.206	14.1	68.4
Sputnik 3	5/15/1958	3.57	1327	372
SCORE	12/18/1958	0.737	15.9	21.6
TIROS	4/1/1960	1.1	122	111
Echo 1	8/12/1960	30.5	66	**2.16**

which resulted in the successful deployment of Echo 1 on orbit. (Figures 1.3 to 1.5 show examples of Echo tests.)

Echo 1 was successfully launched on August 12, 1960, aboard a Delta rocket to an initial orbit of 1600 km (1000 miles), which changed over a period of months from the effects of solar pressure. Echo 1 was operational for a number of months, which indicated that it maintained a sufficiently large and reflective profile for this time period.

This important technology development was significant for a number of reasons:

- Large-area thin films were handled, processed and fabricated.
- Large-size thin films were assembled with high precision.
- The mechanical packaging was highly efficient.
- There was residual gas evacuation and on-orbit deployment by inflation with sublimating powders.
- Launch restraint and orbital release techniques were used.
- Thin films underwent a metallization process.
- Echo was the first large and high-precision operational inflated space structure on orbit.

Fig. 1.2 The first pass of Echo 1. (Photo courtesy NASA.)

Fig. 1.3 Echo subscale pressure test. (Photo courtesy NASA.)

The challenges that arose during the development of this new technology are explained no better than by one of the principal engineers on the project, Edwin Kilgore:

> KILGORE: Yes. Well, as I said, I developed the first sub-satellite using this Mylar material, and the problem was getting it scaled up to 100 feet diameter and put into a small ball. I remember the problem we were having with folding it. If you ever tried to take several acres of Mylar, which is kind of flimsy, and fold it, it just wants to go here and there and everywhere. So it was a real problem.
>
> I was home one rainy morning, and I was getting ready to go to work, and my wife took out her rain hat. You've seen rain hats, you just flip, and they come into a very neat [strip], and you unfold them and they come into a hemisphere [to fit] your head. I said, "I've got to have that." She says, "I need it." I said, "No, I've got to have it worse than you." So I took this out and I gave it to one of the guys down at the shop, and I said, "Look, you go out to the company that's building this Echo balloon, and you take this and you make some forms out of plywood, and make it a real orderly process so that it will look just like this rain hat." That's the way that the thing got folded.

Fig. 1.4 Echo static inflation test. (Photo courtesy NASA.)

Fig. 1.5 LaRC engineers Edwin Kilgore (center), Norman Crabill (right), and an unidentified man take a peek inside the vast balloon during inflation tests. (Photo courtesy NASA.)

> We had a trial balloon, so to speak, in which we fired a rocket going up and then the rocket coming back down, and on the way down it released the capsule and the pyrotechnics fired to open the capsule, and the balloon came out. But what came out were just 10,000 pieces of confetti. It had all this residual air in it, [which] just blew it all to smithereens. It doesn't take much in a vacuum to blow things up, and this, of course, is [a] fragile material anyway. So we had to get the residual air out of it, and that's the reason that it's packed in such a tight space [7].

Residual air entrainment is only one example of the new design and fabrication challenges arising from this new class of spacecraft:

> When the satelloon [the Echo satellite balloon] container is opened to release the satelloon in the hard vacuum of space, any air inside the folded satelloon or outside of the satelloon between its . . . folds tends to expand with explosive rapidity and rip the satelloon to pieces. But this understanding of the problem was not easily acquired, for there is no vacuum chamber on earth big enough and capable of attaining the hard vacuum of space, in which the ejection and complete inflation of the satelloon could be performed and the process photographed with high speed cameras to detect malfunctionings of the process.
>
> Before risking the launch of a balloon into space, the Project Echo Task Group determined that it should first conduct a static inflation test on the ground to see whether the 100 foot-diameter satelloon would assume a spherical shape with surface conditions sufficient to serve as a passive communications relay satellite between two distant stations on the surface of the earth. To make the static inflation tests, Jesse Mitchell took a team of engineers to nearby Weeksville, North Carolina, off the north shore of the Albemarle Sound, where a cavernous navy blimp hangar big enough to inflate the Echo balloon to full size stood empty. The inflation process was slow, taking more than 12 hours, and thus did not offer a dynamic simulation of the explosive inflation that would take place in space; however, the results did reassure everyone that the balloon would work as a communication relay. As Norm Crabill, present at the Weeksville tests, explains, "It was another one of the tests we had to go through before we could trust the design." These tests also demonstrated that the original balloon, manufactured by General Mills, was seriously defective. When the balloon was inflated in the hangar, the triangular panels, or gores, began coming apart at the seams [8].

The modern realization of O'Sullivan's 1956 concept, through experiment, test, and demonstration, is contained in the rest of this book. (O'Sullivan became one of NASA's most highly publicized scientists. In December 1960, the U.S. Post Office issued a commemorative 4-cent stamp in honor of his beloved Echo balloon. For his concept of the inflatable space vehicle, NASA awarded him one of its distinguished service medals.)

1.5 WHERE WILL IT END?

As you read this book, you will realize the great progress that has been made in the testing of LULS. However, many challenges remain.

Fig. 1.6 A 20 m solar sail and boom system developed by Orbital ATK Space Systems of Goleta, Calif., is fully deployed during testing at NASA Glenn Research Center's Plum Brook facility in Sandusky, Ohio. (Photo courtesy NASA.)

A fundamental but critical challenge is the adequacy of the test facility. For example, the NASA Plum Brook Space Power Facility in Sandusky, Ohio, houses the largest vacuum chamber in the world; it has a diameter of 30.5 m (100 ft) and is 36.6 m (120 ft) tall. Figure 1.6 shows a 20 m (on a side) solar sail engineering development model in the chamber; the sail (with a 28.3 m diagonal) almost fills the floor space. Yet the smallest practical solar sail may be about 70 m on a side, and that is impossible to vacuum test today.

One way or another, these hurdles will be crossed, because the grandness of space—and the grand craft that travel within it—embolden grand visions. Here is what President Eisenhower said upon the 1960 launch of Echo 1:

> This is one more significant step in the United States' program of space research and exploration being carried forward for peaceful purposes. The satellite balloon, which has reflected these words, may be used freely by any nation for similar experiments in its own interest [8].

That was a grand idea.

REFERENCES

[1] Jenkins, C. H., Ed., *Gossamer Spacecraft: Membrane/Inflatable Structure Technology for Space Applications*, American Institute of Aeronautics and Astronautics, Reston, VA, 2001, p. 191.

[2] Banik, J. A., "Realizing Large Structures in Space," *The Bridge*, Vol. 45, No. 4, 2015, pp. 22–27.

[3] Collins, H. M., "Public Experiments and Displays of Virtuosity: The Core-Set Revisited," *Social Studies of Science*, Vol. 18, No. 4, 1988, pp. 725–748.

[4] Kass, R. A., "Tests and Experiments: Similarities and Differences," *ITEA Journal*, Vol. 29, 2008, pp. 294–300.

[5] Elder, D. C., "Something of Value: Echo and the Beginnings of Satellite Communications," *Beyond the Ionosphere*, NASA SP-4217, Chapter 4, Part II, 1997.

[6] O'Sullivan, W. J., Jr., interview with Edward Morse, Aug. 28, 1964, "Historical Origins of Echo I," NASA History Office.

[7] Kilgore, E. C., interview with Sandra Johnson, May 3, 2008, NASA History Office.

[8] Hanson, J. R., "The Odyssey of Project Echo," *Spaceflight Revolution*, NASA SP-4308, Chapter 6, 1995.

CHAPTER 2

Structure Classifications and Test Philosophy

Phil Henderson*
Palm Bay, Florida

2.1 INTRODUCTION

Although this book focuses on spacecraft testing that is in some way unique to large ultra-lightweight structures, this chapter also lists those tests considered conventional and necessary for all spacecraft structures. However, in subsequent chapters, further discussion of these standard tests is limited to tests that must be conducted in an unconventional way due to the unique nature of ultra-lightweight structures (such as deployment repeatability and accuracy with sensitive gravity compensation). Also, this chapter does not cover material property or component testing and characterizations, unless they are particularly relevant to large ultra-lightweight spacecraft (LULS) and structures. Numerous standards and references thoroughly cover requirements and guidance for conventional testing of spacecraft structures [1–9].

This chapter is not intended to provide a regimented list of tests that must be done for every type of ultra-lightweight space structure. Rather, the chapter provides a means to evaluate the characteristics of the spacecraft structure and points to key elements to consider when developing an overall test plan. Test philosophy determines what is important and what compromises may be acceptable to achieve the project's objectives. A test philosophy discussion here can point only to various considerations and not to a one-test-plan-fits-all specification. Instead, the chapter provides guidance for making intelligent choices of what ultra-lightweight spacecraft structures to test, and how.

Spacecraft structures can be classified by level of assembly, structural type, critical dimensions, and measurement precision requirements; all of these might apply to conventional spacecraft structures. As we look at classification having to do with critical dimensions and dimensional precision, we start to see the many examples of lightweight spacecraft structures, the focus of this book. Later, we discuss the test philosophy, including the purpose for testing, considerations for an appropriate testing philosophy, and types of tests that are specifically relevant to ultra-lightweight spacecraft.

*Harris Corporation

This chapter discusses

- Classification of structures
- Purposes for testing
- Considerations of appropriate test philosophy
- Appropriate tests for space system structures

A significant truth learned early in the author's career is "There is no substitute for a good test," with the corollary "There is nothing worse than a bad test." The purpose of this book is to be a guide to selecting good tests and a good test philosophy.

2.2 CLASSIFICATION OF STRUCTURES

A *structure* is defined here as a physical object that is capable of carrying loads, usually from one interface to another.

A structure may be classified by the level of assembly, structural/material type, principal critical dimensions, and dimensional precision and stability. Each classification determines the testing approach, as shown in Table 2.1. Further explanation and examples follow.

2.2.1 LEVEL OF ASSEMBLY

Formal testing requirements for conventional spacecraft are usually specified according to level of assembly (as shown in Table 2.1) and are described in MIL-STD-1540 [7]. Ultra-lightweight spacecraft may need to satisfy high-level system specifications or possibly demonstrate why the normal test requirements are not appropriate. Conversely, more appropriate tests for ultra-lightweight

TABLE 2.1 STRUCTURAL CLASSIFICATIONS AND TESTING APPROACHES

Structural Classification	Testing Approach
Level of assembly	Selection of verification tests at the lowest level of assembly in order to reduce risk
Structural/material type	Identification of specialized materials or structural configurations that have established standard requirements or tests
Principal critical dimensions	Determination of complexity of required measurements
Dimensional precision and stability	Determination of measurement precision, need for repeatability tests, or need for measurement over time

spacecraft structures would be added to the "standard verification test" set. The classifications by level of assembly are

- Material tests (such as composite lamina, wire, and thin film)
- Component tests (such as composite laminate, woven mesh, and solar sail panel)
- Subassembly or subsystem (such as truss structure and deployment mechanism)
- Assembly, system, and unit (such as unfurlable antenna reflector)

This book concentrates on tests that are unique or that require unique environments or test methods for large ultra-lightweight structures.

2.2.2 STRUCTURAL/MATERIAL TYPE

Standard testing and verification specifications are also organized by structural or material type, and these types may require specific related verification tests. For example, MIL-STD, European Space Agency (ESA), and NASA test method specifications are available for most all unique structural and material types [1–4]. A few of the common structural and material types are

- Metals
- Polymer films
- Composite laminates and tubes
- Composite sandwich panels
- Foams
- Mechanisms
- Inflatables
- Membranes, flat
- Membranes, shaped
- Wires and cable
- Additive manufactured parts (possibly for all of the above)

Note that there can be interconnections where a structural type, such as an inflatable, can also be a shaped membrane made of a polymer or metallic film, each having some unique tests specified by the standards documents.

2.2.3 PRINCIPAL CRITICAL DIMENSION(S)

Ultra-lightweight spacecraft and structures usually have performance requirements that involve achieving and/or maintaining one or more critical dimensions,

TABLE 2.2 CLASSIFICATION BY CRITICAL DIMENSION OF STRUCTURE

Critical Dimensions	Examples
One dimensional	• Tethers • Control tensioners • Struts and masts (with length requirements only)
Two dimensional	• Struts and masts (with bending and/or twist requirements) • Solar sails • Solar energy (photovoltaic) arrays • Solar shields • Solar and stellar occulters
Three dimensional	• Solar concentrators • Radio Frequency (RF) reflector antennas • Phased arrays • Fresnel membranes (when flatness matters) • Optical telescopes • Optical benches • Optical baffles • Bus and supporting structures

as shown in Table 2.2. The significance or tolerance of a critical dimension varies with the application. For example, a solar array might tolerate some small angles of twist but ultimately must remain pointed to the sun with some defined tolerance. On the other hand, a solar concentrator for a solar dynamic power system (i.e., a closed thermodynamic cycle with a solar-heated working fluid) might have extremely tight pointing requirements to prevent the focal point from wandering. As a result, a hot spot could be created, critically overloading the receiver's heat exchanger.

2.2.4 DIMENSIONAL PRECISION AND STABILITY

Recognition of what is critically important regarding the precision and stability of the assembly leads to the selection of appropriate test methods. Table 2.3 shows classification of structures by dimensional precision and stability.

Finding the right balance of measurement accuracy versus cost of a test is an important aspect of the test philosophy. Identification of the level of precision can dictate the approach needed (and the associated complexity and cost) for various stages of testing. For example, the type of structure may determine the criticality of offloading accuracy and the required complexity of the counterbalance system. However, for high-precision structures, deployment testing may require only offloading that neither impedes nor helps with the deployment motion. Once the test article is deployed, an accurate measurement of a critical dimension might require

TABLE 2.3 CLASSIFICATION BY DIMENSIONAL PRECISION AND STABILITY

Precision and Stability	Examples
Least precision with minimal dimensional stability requirements	• Sunshields • Solar arrays • Solar sails • Tethers
Moderate precision and stability	• Deployable antennas (frequency <100 GHz) • Fresnel lens • Solar concentrators • Struts • Mast
Extreme precision and stability	• Telescope systems (wavelength $<10^{-3}$ m) • Stellar occulters • Optical benches

a completely different test fixture approach or a different level of offloading accuracy or vibration isolation from ground sources. In another example, an assembly that spins during deployment might require counterbalancing as well as a means to counteract air damping during deployment. In this case, deployment testing might be accomplished inside a rotating room (with a rotating air mass) rather than in vacuum or low-pressure chamber. Later, this same assembly may not require the spinning room for the precision measurement of the deployed reflector surface (if the operational rotation on-orbit will not distort the surface).

2.2.5 METHOD OF ACHIEVING STIFFNESS

Ultra-lightweight structures are more likely to be tension-stiffened than bending-stiffened because the former can outperform (i.e., stiffness-to-weight) the latter by orders of magnitude. This differentiation in structural design and associated testing is discussed in Chapter 3 and Chapter 6.

2.3 PURPOSES FOR TESTING

Tests are conducted for many reasons, including risk reduction, engineering development, concept validation, validation of analysis models, verification of function or performance, demonstration of interface compatibility, evaluation of failure modes, and in combination with analysis, for example, to verify compliance.

2.3.1 RISK REDUCTION

Risk reduction is a common reason to conduct tests throughout both the design and manufacturing processes. All space systems are high-value projects. Deployable ultra-lightweight structures are likely to be an enabling part of a larger space system such as antennas or solar arrays. Space hardware customers and suppliers invest significant time, money, and reputation in their projects. A failure in orbit is usually difficult to troubleshoot and to recover from, highly visible to the public, and not soon forgotten. Although a lot of risk-taking may be expected with "New Space" projects (flying a low-cost ride to orbit and a CubeSat bus), aerospace companies (large and small) are just as interested in guaranteeing success of the CubeSat project as they are on a very high-profile mission such as a Mars lander. Thus, risk is relative.

Each project must have an evaluation of the cost and schedule for conducting a particular test compared to the risk of an unexpected failure later in manufacturing, or in a ground test at a higher assembly, or in orbit. Conventional testing handbooks give some guidance [5–9] based on long-accepted practice. Many space customer organizations have required standards or a process to obtain a waiver. Often, customer standards for testing are derived from heritage processes and take precedence over what is actually needed. Conversely, the ultra-lightweight spacecraft engineering team should have a much better understanding of the additional risk-reduction tests that should be conducted beyond what is specified. Do not overlook the importance of engaging specialists with experience to assess risks and perform tests to assist planning and execution.

2.3.2 ENGINEERING DEVELOPMENT

Development tests early in the concept phase of the project may validate an approach. The test may be accomplished faster and more cost effectively than detailed analysis. Sometimes, designs cannot be analyzed adequately due to lack of material or assembly properties. A quick test may be enough to make concept trades and design decisions.

2.3.3 CONCEPT VALIDATION

Validation testing can extend from the initial proof of concept to the final verification before delivery of the completed assembly.

2.3.4 VALIDATION OF ANALYSIS MODELS

Validation testing with correlation to analysis models allows extrapolation from the test environment into operational environments that cannot be simulated on the ground. When sufficient accurate analysis is feasible, but testing in the complete operational environment is costly or not feasible, analytical extrapolation is useful. Extrapolation carries some risk, but sometimes, there is no other option. Discernment is usually necessary, such as when the assumed extremes of the environment are tested and environments between the extremes are not tested. Again, risk-versus-reward assessment justifies this interpolation.

2.3.5 VERIFICATION OF FUNCTION OR PERFORMANCE

Verification tests show the compliance to requirements specifications. Such tests include

- Function tests (such as deployment kinematics and life)
- Critical performance (such as antenna surface precision)
- Material or component limits (i.e., qualification)
- Survival and operation in specified environments
- Interface compatibilities (such as interface loads, circuit connections, and grounding)

2.3.6 IN-ORBIT VERIFICATION

In-orbit verification tests of structures manufactured or assembled in space may be necessary if the fully deployed assembly cannot be adequately tested on the ground. Very large gossamer structures, even with the most careful offloading, behave differently in the actual orbital environment. Even then, a subscale test will likely be done on the ground to demonstrate or validate the structure before fabrication of the full-scale project in orbit. See Chapter 7 for more about spaceflight testing.

2.3.7 EVALUATION OF FAILURE MODES

Identification of failure modes leads to prevention and mitigation actions, or later, if not successful, identification of the cause of a failure.

2.3.8 COMPLIANCE VERIFICATION BY COMPLEMENTARY ANALYSIS AND TEST

Complementary tests combined with analysis may accomplish what cannot be done by one method alone. For example, end-to-end alignment of an ultra-lightweight structure might not be possible while it is attached to the spacecraft on the ground with (1g) gravity loads. Limitations of spacecraft orientation and gravity-offloading might result in significant risk to the hardware or might cause large measurement uncertainty. Alignment measurements of separate segments of the ultra-lightweight structure, plus finite element model deflection analysis and coordinate system transformations, enable prediction of the zero-g alignment to the spacecraft. Similarly, the combination of test measurement and analysis provides a practical means to predict antenna performance (by measuring the primary pattern of the feed and analytically combining that with a measured reflector surface contour), or for predicting mass properties (by weighing individual components and combining them in the Finite Element Model [FEM] or Computer Aided Design [CAD] model).

2.4 CONSIDERATIONS OF AN APPROPRIATE TEST PHILOSOPHY

Appropriate test philosophy considers the type of LULS structure, purposes for testing, and the minimum testing necessary to

1. Validate a new concept before full-scale development.
2. Reduce risk at the lowest possible level of assembly.
3. Verify the intended performance or capability to withstand environments.
4. Show compliance to a specification.

Ultimately, each hardware development team and its customer must decide the appropriate tests for the project to ensure a successful mission in space.

Numerous standards and specifications cover the testing of conventional spacecraft structures. LULS structures present special challenges to satisfy the standard requirements, while accounting for the special properties of the ultra-lightweight (such as ultra-low areal density, flexure-stiffened, tension-stiffened, folding to fit the launch configuration, not self-supporting in 1g, or needing counterbalance in ground performance tests). Additional considerations influence the appropriate testing philosophy.

2.4.1 SEQUENCE OF TESTING

Sequence of testing is as important with ultra-lightweight structures as with conventional spacecraft structures. However, the proper sequence may depend on the failure mode. Some examples are

- Payload testing specifications normally require all structural load and dynamic tests to take place before powered operation in thermal vacuum because certain structure failure modes (such as bad solder joints, broken wires, and poor welds) become detectable in the extreme thermal environment.
- Conversely, space structure thermal cycle tests are run before vibration tests and load tests to demonstrate that the thermally induced micro-cracks do not reduce material allowables to the extent that the microscopic flaw leads to catastrophic structural failure during launch.
- Sine vibration is run before random vibration, and then, afterwards a sine sweep is done to uncover structural response changes.
- A phenomenon of RF antennas is called Passive Inter-Modulation (PIM), where unwanted signals generated by material assembly non-linarites can cause severe

signal noise and sometime self-jam a communications satellite. An antenna must be shown to be "PIM free" (i.e., PIM signals are below a defined threshold) after all environmental tests are complete as a final check before launch.

The bottom line: Consider the sequence of the tests relative to failure generation and failure mode detection for the specific structural assembly.

2.4.2 INSPECTIONS AND MEASUREMENTS

Inspections and measurements are necessary before and after each stage of testing to determine the structure's stability and to satisfy pass/fail criteria. An appropriate test philosophy has at least some "health check" measurements after each test in a series, even if full measurements are not feasible. Thus, if a failure or change occurs, it is apparent which environment caused the change. Ultra-lightweight structures may require special test tooling and methods for these measurements. Gravity compensation and deployed position repeatability are covered in Chapters 3 and 5, respectively.

2.4.3 COMMERCIAL VS GOVERNMENT PRACTICE

Commercial practice versus government practice frequently demonstrates a disparity between what testing is required by commercial satellite customers and what is required by government space system programs. Commercial customers are more risk averse because survival of the business depends on the success of the project. It is often thought that commercial projects take more risks, probably because the business plan drives the schedule urgency of commercial programs more often than is evident in the long development cycles of government procurement. Commercial programs nearly always require immediate application of flight-proven, off-the-shelf hardware versus a more leisurely progressive series of studies, hardware development, and schedules driven by funding limits and fiscal years, typical of government-funded projects. Oversight of commercial programs focuses on performance, schedule risk, and milestone achievement versus the government emphasis on technology development and process, driven by the oversight review of multiple agencies and consultant groups. As the space industry has grown in recent years and funding becomes scarcer, more and more government projects adopt commercial approaches. Conflict can occur when the government attempts commercial-like procurement but continues the oversight practices that drive up performance and increase requirements, and impact both schedule and cost. Therefore, appropriate testing for technology development can be much more extensive than testing needed for commercial applications.

2.4.4 NEW VS CONVENTIONAL SPACE

New space versus conventional space regarding spacecraft design and testing refers to a lower-cost approach, an innovative use of off-the-shelf technology, or possibly an incremental development process with flight systems initially having lesser complexity and capability than later improved versions. Easier access to orbit enables new-space thinking. Rather than handbook guidance or one-size-fits-all test methods, the new-space approach requires creative use of available resources and empowerment to suggest what is appropriate for the specific ultra-lightweight spacecraft structure. However, do not neglect the previous discussion regarding risk with new-space projects. Testing appropriately before flight is always better than inappropriate risk taking and failure on orbit.

2.4.5 STANDARD TEST STAGES

Table 2.4 shows standard test stages for typical space systems projects that incrementally reduce risk, then validate and verify the large ultra-lightweight structure.

2.4.5.1 DEVELOPMENT TESTS

Development tests occur very early in the design process. These tests are typically quick and informal, but they should be controlled and documented. Development tests are for learning, they may be iterative, and they often provide input for trade studies between competing designs. Such tests are very appropriate for new space structures and mechanisms.

2.4.5.2 QUALIFICATION TESTS

Qualification tests cover all levels, including components, subassemblies, and the completed top assembly (spacecraft). The qualification test builds confidence, confirms design selection, and establishes margins in conditions more extreme

TABLE 2.4 PURPOSE FOR TESTING DURING DIFFERENT PROJECT STAGES

Project Test Stage in Sequence	Development	Risk Reduction	Validation and/ or Correlation	Identification of Failure Modes	Complement Analysis	Verification
Development test	x	x		x	x	
Qualification test			x	x	x	x
Protoflight test			x		x	x
Acceptance test					x	x
Flight test	x		x		x	x

than those expected in flight. Usually, the qualification test is accomplished with a dedicated unit that increases costs but provides a thorough means of establishing environmental margins. The dedicated qualification unit may be capable of being refurbished and return as a flight unit after acceptance testing. Sometimes a development test may be elevated to a qualification test at the component level or at lower levels of assembly. If so, the development test article must be appropriately similar to the intended flight article. Appropriate similarity can be subjective and is the key topic discussed at a space project Equipment Qualification Status Review.

2.4.5.3 PROTOFLIGHT TESTS

The protoflight test approach utilizes the first flight unit rather than prior testing with a dedicated qualification unit and may be justified for several reasons. Protoflight testing uses assemblies comprised of previously qualified components and qualified subassemblies when it is accepted that the final unit assembly will have no effect on the individual lower assembly qualifications. Further justification considers the sum of the parts as substantially similar to another qualified assembly, or extensive previous flight heritage of similar designs.

The protoflight approach is normally employed when time or funds are limited, such as when there is no time to build and test a qualification unit in series with the flight units. The Protoflight approach is also used when margin requirements are low and environments are well known, or with workmanship-dependent assembles. In the latter case, each unit in a series might be exposed to protoflight testing levels to make sure there is sufficient margin between the test levels and the expected worst case in flight. Many LULS structures programs have successfully used a protoflight approach for both government and commercial programs.

Protoflight vibration tests are conducted at the same test levels as in qualification vibration environments but for shorter durations. Protoflight thermal tests are run to temperature levels not as extreme as qualification levels.

2.4.5.4 ACCEPTANCE TESTS

Acceptance test levels and durations verify normal workmanship and are used on flight units following the dedicated qualification unit or protoflight unit. An acceptance test program typically verifies specific, but not all, environmental and performance margins. One very successful flight program with unfurlable mesh antenna reflectors (NASA's original Tracking and Data Relay Satellite [TDRS]) flew a dozen units with only a workmanship low-level vibration acceptance test after a successful qualification test of the first unit.

2.4.5.5 FLIGHT TESTS

When the spacecraft structure cannot be tested adequately on the ground, flight tests may be the only way to demonstrate full function or on-orbit adjustment, or to measure on-orbit dimensional accuracy and stability. In the case of CubeSat programs, a series of flight tests may really be a series of on-orbit development tests where successive flights discover and eliminate design flaws.

2.5 APPROPRIATE TESTS FOR SPACE SYSTEM STRUCTURES

As we have stated, ultra-lightweight spacecraft structure tests are often conventional tests with special considerations for gravity offloading, precision measurement over large dimensions, special test fixtures, and more. The tests listed here are appropriate for ultra-lightweight spacecraft structures and are typical for all spacecraft structures. They are listed with the intent to be inclusive and, at the same time, to identify the special considerations covered in more detail later in this book.

Understanding the level of assembly, structural type, principal critical dimensions, required accuracy, and the stage of the project helps identify the necessary tests and methods. Table 2.5 shows the various structural types mapped to tests discussed throughout the book.

Special performance tests (such as the antenna PIM screening, shown in Fig. 2.1) may be required for specific LULS structures (although not included here) and may require offloading if the structure is not self-supporting in 1*g* or if surface accuracy is required.

2.5.1 ENVIRONMENTAL EXPOSURE FOR SPACE SYSTEMS

Verification of function, performance, and durability during and after relevant environmental exposure are the "main events" of spacecraft structural testing. These topics represent an initial list, not the total of the considerations to be made.

2.5.1.1 DEPRESSURIZATION TESTS

Depressurization tests demonstrate adequate venting (i.e., what happens to the gas volume inside enclosures during the Boyle's Law gas expansion at launch). Such tests are critically important for packed membranes and any other enclosures. Solar sails, thin folded or rolled membranes, large multi-layer insulation blankets, or any other large structure that can trap gas and become pressurized require appropriate testing and possibly special design consideration, as well as on-orbit deployment procedures that can compensate for entrapped gas issues. For example, the JPL Large Inflatable Deployable Antenna Flight Experiment (IAE) discovered that residual air in the inflatable structure prematurely deployed the structure when the launch container was opened, causing some risky unplanned deployment kinematics [10].

TABLE 2.5 SPECIAL TESTS OR TEST TECHNIQUES NEEDED FOR LULS STRUCTURES

	1-D			2-D				3-D								
	Tethers	Control Tensioners	Struts and Masts	Solar Sails	Solar Arrays	Solar Shields	Occulters	Solar Concentrators	RF Reflector Antennas	Phased Arrays	Fresnel Membranes	Optical Telescopes	Optical Benches	Optical Baffles	Bus and Supporting Structures	Comments and Handbook References
Dimensional Precision and Stability																
Least	x			x	x	x								x		
Moderate		x	x					x	x	x	x				x	
Extreme							x					x	x			
Environmental Tests																
Depressurization				x	x	x		x	x		x					Assemblies with folded membranes where entrapped air may impact deployment reliability
Thermal Vacuum/ Solar Thermal Vacuum					x			x	x	x	x	x	x			For vacuum-sensitive assemblies or where thermal gradients impacting surface accuracy can only be simulated in a solar thermal environment
Thermal Cycle		x	x				x		x	x		x	x			Thermal cycle should be done for any dimensionally critical components prior to performance testing.
Heat Balance										x			x			For verification of active thermal control on dimensionally critical structures
Static Loads		x	x				x		x	x		x	x			Static loads (or quasi-static load vibration tests) should be done for any dimensionally critical components prior to performance testing.
Dynamic Loads		x	x				x		x	x		x	x			Dynamic load tests should be done for any dimensionally critical components prior to performance testing if not enveloped by static load tests.
Material Survivability	x	x	x	x	x	x	x	x	x	x	x	x	x	x	x	
Functional Tests																
Deployment	x	x	x	x	x	x	x	x	x	x	x	x	x	x	x	Offloading may be required. See Chapter 3, Gravity Compensation.
Unfurling Membranes				x	x	x	x	x	x		x					Offloading may be required. See Chapter 3, Gravity Compensation.
Inflation (if inflatable)			x					x	x	x					x	Offloading and pressure differential issues See Chapter 3, Gravity Compensation.
Mechanism Operation or Demonstrations	x		x	x	x	x	x	x	x	x	x	x			x	Offloading may be required. See Chapter 3, Gravity Compensation.
Life Testing																
Electrical Connectivity																
ESD Grounding																
EMI/EMC																
Telemetry and Sensor Validation																
Performance Tests																
Stowed Stiffness																
Deployed Stiffness			x				x	x	x	x	x	x	x			Gravity compensation required and may interact with measurement See Chapter 3, Gravity Compensation and Chapter 6, Structural Dynamics
Deployed Accuracy	x	x	x				x	x	x	x	x	x	x			Measurement technique and gravity offloading issues. See Chapter 3, Gravity Compensation and Chapter 5, Deployment Repeatability
Thermal Distortion			x				x	x	x	x	x	x	x			Gravity compensation required and may interact with distortion measurement. See Chapter 3, Gravity Compensation
Alignment			x				x	x	x	x	x	x	x			End-to-end alignment may require combination of measurement of individual assemblies, combined (coordinate transformations) with analytical techniques
Mass Properties									x			x				Inertia tests or spin balancing may require special handling or tests

Fig. 2.1 This unfurlable mesh antenna PIM test did not require offloading for precision but only for unloading the deployable boom. (Photo © Harris Corporation 2016. Used with permission.)

2.5.1.2 THERMAL VACUUM/THERMAL CYCLE TESTS

Thermal vacuum/thermal cycle tests verify survival and performance in the expected temperature environment.

- Full thermal vacuum testing is appropriate if bearing surfaces, vacuum-sensitive restraint release, or mechanism operation cannot be tested adequately at the lower assembly level.
- Thermal cycle tests (in an inert atmosphere) may be an appropriate substitute for thermal vacuum tests on some ultra-lightweight spacecraft structures when it is shown that there are no vacuum-sensitive components such as bearing surfaces, thermal systems, or pressure sensitive components. If there are vacuum-sensitive subassemblies, an appropriate thermal vacuum test can be conducted with the component at a lower level of assembly. This combined test approach also allows for faster and more focused verification tests since components can be tested in a wider temperature range than would be permitted with the full assembly due to limits on another part of the assembly.

- For assessing performance of extreme accuracy surfaces with a realistic temperature distribution, consider a solar thermal vacuum (STV) test. Figure 2.2 shows an unfurlable mesh reflector antenna in an STV test. Once thermal and thermoelastic finite element models have been correlated in the realistic environment, the performance can be determined analytically for follow-on similar configurations, without the need for the usually expensive thermal vacuum test.

2.5.1.3 HEAT BALANCE/HEAT LEAK TESTS

Heat balance or heat leak tests of multi-layer insulation (MLI) thermal control blankets over dimensionally critical space structures are appropriate to evaluate MLI heat transfer properties, an important modeling input to the temperature prediction analysis that results in a more accurate structural distortion prediction.

- Whenever a new configuration of MLI blanket or new thermal control is developed, an assessment must be made to determine if previous tests and thermal properties are still valid. If not, an updated heat leak test may be necessary. Space system contractors frequently maintain a design thermal manual with MLI blanket and thermal control properties. However, the application of the

Fig. 2.2 Unfurlable mesh reflector undergoes solar thermal vacuum test. (Photo © Harris Corporation 2016. Used with permission.)

property to a real and complex assembly may be susceptible to modeling errors. Heat leak tests of the individual subassemblies with all thermal controls in place provide a more accurate input for the thermal model of the full assembly and are consequently a better prediction of temperatures and thermal distortion.

- Results of separate heat leak tests of subassemblies can be combined analytically into the full assembly thermal model, much more feasible than one heat leak test of the full assembly.
- Sensitivity analysis that evaluates thermal property uncertainties may show that large uncertainty has little effect on thermal distortion of the full system. Surface performance budgets may also show the relative importance of the nominal prediction relative to the uncertainties and indicate whether additional thermal property testing is needed to reduce distortion uncertainty.

2.5.1.4 STATIC LOADS

Static loads can be accomplished by proof-loading with a static test fixture if the structure has sufficient interface points for adequately loading. Static loads may be associated with the launch, deployment, or some other on-orbit condition such as spacecraft maneuvers. Examples of the static load test are a bending or twist test of a beam or truss structure.

- Proper orientation with the gravity vector and offloading may be required for ultra-lightweight structures.
- "Quasi-static loads" are often specified to envelope the dynamic loads of launch. Static testing to the quasi-static load levels verifies the structure's ability to withstand dynamic load testing.
- Large deployable structures are often static load tested one component at a time to verify bonds and mechanically stabilize final dimensions.

2.5.1.5 DYNAMIC LOADS

Dynamic loads include sinusoidal (sine) and random vibration, acoustic input, and shock loads. Acoustic and shock tests are launch loads considered conventional and covered in other references [1, 5–9, 12].

- The usual issue with sine vibration loading involves the specification of acceptable test level input limitations around structural resonant frequencies. Input limits, or "notching," at the assembly interface mounts prevent over-stressing of one portion of the structure. Large ultra-lightweight structures with multiple structural interfaces to the spacecraft often require this kind of test approach. Sine vibration and acoustics may both require the notching approach to limit resonant loads, as described earlier.

- Over testing the assembly is sometimes considered if the component that limits notching has a large strength margin. This over testing can be risky due to incomplete knowledge from modeling and subassembly tests.
- Modal surveys before and after shock or vibration loading uncover changes resulting from damage during the test.
- Structural dynamics for ultra-lightweight structures are covered in Chapter 6.

2.5.1.6 ACOUSTIC LOADS

Acoustic loads are often the most severe load environment for large lightweight structures but they also affect other components that respond to the primary structural vibration, such as insulation blankets, secondary structures, and cables.

- Acoustic load testing of large lightweight structures that stow on the outside of a spacecraft should be done on either the spacecraft or on an accurately simulated spacecraft mounting surface so that input loads (amplification and damping) are as realistic as possible.
- Usually a random vibration test can be substituted for the acoustic test if the large ultra-lightweight structure does not respond to acoustic loads. This is true for stowed mesh antennas since even large areas of folded mesh do not support a pressure gradient. Thus, the acoustic environment does not cause any additional structural load to the spacecraft other than the mesh antenna's dynamic response to the interface loads generated by the spacecraft panel.

2.5.1.7 MATERIAL SURVIVABILITY TESTS

Material survivability tests may be appropriate for some types of structures and environments. Table 2.6 shows material tests that may be applicable to large ultra-lightweight space structures.

- Combined effects testing may be key in multiple simulated environments. Such testing can be costly but has the potential to save the mission. With the Hubble telescope MLI blanket, for example, although every environment was tested separately and passed, insulation blanket deterioration occurred on orbit. Ultraviolet (UV) radiation, combined with high temperature extremes in the actual orbital environment, breaks down polymer chains, causing the MLI blanket's aluminized-Teflon® fluorinated ethylene propylene (Al-FEP) outer layer to become much more embrittled. Also, combined UV radiation and Atomic Oxygen (AO) exposure resulted in higher erosion rates than with AO alone [11].
- Kapton, a polyimide film used frequently for multi-layer insulation blankets, is particularly sensitive to AO.

TABLE 2.6 SURVIVABILITY TESTS FOR SOME TYPES OF STRUCTURAL MATERIALS AND ENVIRONMENTS

Survivability Test	Material or Environment
Vacuum exposure outgassing	For composites and polymers.
Radiation (particle, UV, etc.)	For materials with critical properties such as tensioned polymers or thermal control coatings that change properties after exposure.
Atomic oxygen (AO)	For low earth orbit (LEO) applications depending upon life requirements and altitude. AO especially affects polymers.
Chemical/ionic environment	For membranes, reflective surfaces, or thermal controls that will be exposed to attitude control thruster products.
Damage tolerance	For structures where handling damage is a factor (such as inflatable habitats).

2.5.2 FUNCTIONAL TESTS

A functional test demonstrates that one or more aspects of a component, subassembly or assembly operates as it is intended for the mission. Functional tests can be as simple as the release of a restraint or as complex as a deployment operation during a solar thermal vacuum test.

2.5.2.1 "TEST WHAT YOU FLY"

"Test what you fly" is the desired norm. Even so, large lightweight space structures cannot always be tested in their fully deployed configuration due to uncompensated gravity effects, safety handling issues, air damping, inflation pressure limitations, test chamber size, and so on. Chapters 3 through 7 address these limitations in greater detail.

2.5.2.2 DEPLOYMENT

Deployment of the ultra-lightweight spacecraft structure is the ultimate test after exposure to all launch environments. Chapters 3, 4, 5, and 7 each discuss specific aspects of deployment testing.

- It has been suggested that a structure is truly launch ready when the test engineer can start the deployment and then leave the room, without fear, while the structure completely deploys by itself using actual or simulated spacecraft controls.

Fig. 2.3 Deployment and surface measurement of large unfurlable mesh reflectors (18-m diameter shown) requires specialized tooling for offloading. (Photo © Harris Corporation 2016. Used with permission.)

- The next best test is a completely hands-off uninterrupted deployment, observed by someone with the "kill switch" in hand. It is recommended that observers look for danger to the test article (or the personnel, of course) and be authorized to yell, "Stop!"
- Offloading and appropriate feedback control telemetry are absolutely necessary. A large deployable mesh antenna with gravity offloading tooling is shown in Fig. 2.3.
- Video, or sequential still photography, is very useful in evaluating the success of deployment on orbit.

2.5.2.3 UNFURLING/UNFOLDING MEMBRANES, MESH, AND CABLES

Unfurling/unfolding membranes, mesh, and cables must deploy in a systematic manner that does not have the potential for hang-ups or snags of the surface (such as wire mesh) or control lines.

- Offloading of the surface may be necessary to prevent gravity from either assisting or resisting the deployment. Gravity compensation is covered in Chapter 3.
- Deployment in multiple gravity orientations (such as cup up or cup down for antenna reflectors) is essential to verify the function of membrane

management devices. Multiple orientations relative to gravity gives the membrane and supporting components a better chance to hang up and helps to identify potential failure modes.

2.5.2.4 INFLATION

Inflation of an inflatable structure is a special case of deployment. Inflation gas can be used to drive the deployment, and then to stabilize the structure.

- Deployment control devices or folding techniques maintain a specific gas pressure in the structure during deployment to provide known stiffness and to guide its deployment path. Without deployment control devices, the structure can deploy in an unpredictable manner and potentially impact the rest of the spacecraft, or become entangled. A simple example of an inflatable device with deployment control is a typical "party favor" that unrolls when blown. Numerous methods of inflation deployment control have been used in spacecraft, including to internal partitions, hook-and-loop tabs, springs, and strain energy devices. All such devices must be thoroughly tested with simulated loads at extreme temperatures.
- Inflatable structures can be especially affected by gravity loads, and adequate off-loading may be very difficult to achieve.

2.5.2.5 MECHANISM OPERATION TESTS

Mechanism operation tests, or demonstrations, occur as part of deployment or restraint release tests and are associated with environmental verification tests.

- Mechanism tests should demonstrate operation in worst-case environments, torque (or deployment) margins, and life (or durability) with appropriately sequenced environments that include manufacturing operations, verification tests and on-orbit operations.
- Restraint release tests should be followed by the initial separation of interfaces, unless the assembly precludes that. For example, in some cases where large multi-segment booms are deployed, only the first set of spacecraft interface restraints are released. Then, removal of the boom assembly may be necessary so that the boom can be fully deployed using floor stands and simple offloading.
- Offloading of the ultra-lightweight structure for the "first motion" test in a worst-case environment (such as in the thermal vacuum chamber) can present a significant design challenge and risk to the hardware, and requires significant attention in test tooling design.
- Individual mechanisms within the large assembly should follow the USAF specification for Moving Mechanical Assemblies [12]. Some tailoring is usually necessary to optimize program cost.

2.5.2.6 LIFE TESTING

Life testing includes both the mechanism tests and material survivability tests described earlier.

- One aspect of life testing is the demonstration of damage tolerance. Evaluation of performance, including precision dimension measurement, can be overwhelmed by damage from handling or repeated loads.
- Damage tolerance for inflatable habitats is more important than dimensional precision or repeatability after life testing.

2.5.2.7 ELECTRICAL CONNECTIVITY

Electrical connectivity verification is critical for deployable structures at the beginning of the deployment verification and in the final tests during integration for launch.

2.5.2.8 GROUNDING

Grounding (such as electrostatic discharge [ESD] prevention) of large deployable structures must be verified to show the specified redundancy in ground paths. Inadequate grounding results in high-voltage discharges that damage wiring, spacecraft components, or membranes. This testing is done directly on the subassemblies and then on the final assembly by ohmic measurement from locations on the structure to the common interface ground points.

- Often, large structures with membranes or tensioned cables must be grounded through the structure since adding dedicated electrical wires may significantly increase the mass.
- Certain techniques to enhance conductivity through bonded joints or across gaps in membrane surfaces result in an adequate ground path.

2.5.2.9 ELECTRO-MAGNETIC INTERFERENCE/ELECTRO-MAGNETIC COMPATIBILITY (EMI/EMC)

EMI/EMC is part of the testing required for electrical motors of drive mechanisms as well as any other active electrical devices. These tests are typically performed in special anechoic test chambers for electromagnetic isolation.

2.5.2.10 TELEMETRY ACTUATION AND SENSOR VALIDATION

Telemetry actuation and sensor validation, usually critical for deployment, should be demonstrated in the lower assembly and full assembly deployment tests. Life testing of the components may be necessary if sufficient vendor test data or heritage is not available.

2.5.3 PERFORMANCE TESTS

Performance tests of ultra-lightweight spacecraft structures are performed in both the launch configuration (stowed) and the fully deployed, operational configuration. Sometimes, performance tests are done during the in-between stages.

2.5.3.1 STOWED STIFFNESS

Stowed stiffness of large deployable structures can be the "tail that wags the dog" on a smaller spacecraft. This test is typically performed using a vibration shaker with sinusoidal sweeps or random vibration input while the structure is mounted on an extremely rigid test tool.

- Some ultra-lightweight structures have multiple stowed structural interfaces to the spacecraft, and vibration coupling with the spacecraft must be carefully evaluated.
- Normally, payload specifications require a relatively high stowed frequency as if an electronic box were being packaged. Large deployable structures may need to be "tuned" so that resonances fall in a range, not just above a lower limit. Otherwise, high stowed stiffness requirements can drive the mass higher and higher during the design phase. Vibration testing of the full stowed assembly with a rigid interface gives confidence in the structural model that can then be employed in coupled loads analysis with the spacecraft and launch vehicle to verify compatibility.
- Coupled loads analysis may be an iterative process to eliminate interdependencies of the spacecraft, the payload, and a large flexible, deployable structure that cannot be predicted.

2.5.3.2 DEPLOYED STIFFNESS

Deployed stiffness affects the design of attitude control systems of spacecraft with large deployable structures. Stiffness affects performance where perturbations of the spacecraft can cause mispointing of precision deployable antennas or optics.

- Deployed stiffness might be verified by modal survey of a supported ultra-lightweight structure.
- Another possibility is a "twang" test, in which an interface is held fixed (with extended structure offloaded) and the unit is perturbed, then released. Deployed stiffness analysis is accomplished using the FEM, with input of measured (appropriately tested) stiffness of joints and structural members and measured weights.
- Some ultra-lightweight structures must be tested in vacuum and in air to calibrate the air-mass damping effects. If the size of the structure is large relative to the available vacuum chambers, subscale testing might be considered.

- Spacecraft control systems have become very tolerant of low-frequency modes and the "tail is wagging the dog." Modern large solar arrays are a good example because the deployed structure modes are well within the control bandwidth. This approach might be applied to other gossamer structures on small but capable spacecraft buses, even high-end CubeSats. Chapter 6 discusses structural dynamics testing in more detail.

2.5.3.3 DEPLOYED ACCURACY

Deployed accuracy defines the utility of many ultra-lightweight structures.

- As an example, unfurlable mesh antenna reflector structural property specifications ultimately relate to antenna RF performance. These properties include as-manufactured surface roughness, surface pointing, and focal length, as well as worst-case distortions in all specified orbital environments over the operating life.
- "As-built" surface accuracy of truly gossamer structures may require special techniques. For example, measurement "cup up and cup down" is used to obtain an average distortion relatable to the zero-*g* environment. Also, offloading with balloons and other devices is usually required to reduce measurement error. Chapters 3 and 5 discuss gravity offloading and deployment repeatability in more detail.

2.5.3.4 THERMAL DISTORTION

Thermal distortion tests can range from the solar thermal vacuum exposure with in-situ surface shape measurement to simpler thermal cycle tests with shape measurement.

- Thermal distortion testing in a solar thermal vacuum chamber achieves the highest gradients and realistic temperatures over a large deployable structure. Thermal vacuum also verifies insulation blankets and heaters that may be critical to on-orbit performance.
- Thermal distortion tests at atmospheric pressure (dry nitrogen) allow for bulk temperature distortion testing. These tests may not reach the extremes of on-orbit temperatures nor important thermal gradients in the structure.
- Component heating is an alternate approach to full assembly test and has proven successful for thermal distortion FEM verification of large deployable antenna structures with tensioned cord trusses. Local heating of isolated structures and simultaneous measurement of the overall assembly distortion provide a verification of performance driving components and subassemblies of the assembly. Correlation of the FEM to the measured structural element distortions gives confidence in the predicted performance of the assembly in other environment cases not tested.

2.5.3.5 ALIGNMENT

Alignment to the spacecraft bus is critical for many deployable spacecraft structures (such as antennas and optical systems) and must be verified. This is also true of structures comprised of multiple subassemblies (such as a deployable boom leading from the spacecraft bus to a large unfurlable antenna reflector).

- End-to-end alignment measurement is not always possible due to the long distance between reference points, or due to errors in 1*g* offloading of large ultra-lightweight structures. For example, alignment verification of a deployable mesh reflector and its deployable boom is accomplished more accurately by measuring the alignment of the interfaces at each end of a boom segment, measuring the parabolic antenna surface relative to the attachment interface, and then combining the alignments analytically in a budget.
- Alignment verification should account for initial accuracy, repeatability, and stability of the deployable structure.
- Alignment to the spacecraft reference coordinate system may require adequate visibility for the reference mirrors and tooling targets to be visible to the theodolite instruments while the assembly is stowed and integrated to the spacecraft. Interface specifications should include requirements for the location and visibility of the targets.

2.5.3.6 MASS PROPERTIES

Mass properties (mass, center of gravity, and moments of inertia) of a deployable ultra-lightweight structure are usually measured directly for the stowed configuration.

- Frequently, center of gravity and moments for stowed assemblies are calculated and tabulated with FEM or CAD models when direct test capability is not available.
- The deployed mass properties may be determined by a combination of mass measurements of lower assemblies, coupled with analytical models, when direct measurement is not feasible.

2.6 CONCLUSIONS

LULS structures have been categorized and compared to determine what types of tests and test techniques may be appropriate for specific physical characteristics.

Appropriate test philosophy begins with the intent to

1. Validate a new concept before full-scale development
2. Reduce risk at the lowest possible level of assembly

3. Verify the intended performance or capability to withstand environments
4. Show compliance to a specification

Some tests within the general categories of environmental exposure, function, and performance may be uniquely required for large ultra-lightweight structures. Other tests considered conventional and necessary for all spacecraft structures must be conducted in an unconventional way for large ultra-lightweight structures.

Although this book can provide suggestions and guidance, each hardware development team (with its customer) has the best knowledge of a particular system and the minimum testing required. The team has the ultimate responsibility to select the appropriate tests for the project to ensure a successful mission in space.

ACKNOWLEDGEMENTS

The author would like to acknowledge the following persons:

- For contributions regarding inflatable structures and review editing: David Cadogen of ILC Dover Corporation
- For review editing: Robert Taylor of Harris Corporation
- For line editing: Joyce Henderson
- For review editing: the book editors, Jeremy Banik, and Chris Jenkins.

Thank you all for the many improvements.

REFERENCES

[1] MIL-STD-810G, *Environmental Engineering Considerations and Laboratory Tests*, Department of Defense Test Methods Standard, 31 Oct. 2008.

[2] ECSS-Q-ST-70-45C, Space Product Assurance, Mechanical testing of metallic materials, 31 July 2008.

[3] ECSS-Q-ST-70-04C, Space Product Assurance, Thermal testing for the evaluation of space materials, processes, mechanical parts and assemblies, 15 Nov. 2008.

[4] NASA-STD-6016, Standard Materials and Process Requirements for Spacecraft, 11 July 2008.

[5] GSFC-STD-1000E, Rules for the Design, Development, Verification and Operation of Flight Systems, NASA Goddard Space Flight Center, 3 Aug. 2009.

[6] GSFC-STD-7000A, General Environmental Verification Standard (GEVS) for GSFC Flight Programs and Projects, NASA Goddard Space Flight Center, 22 April 2013.

[7] MIL-STD-1540, *Test Requirements for Launch, Upper-Stage, and Space Vehicles*, Department of Defense Standard, 15 Sept. 1994.

[8] MSFC-HDBK-2221, *Verification Handbook, Volume 1: Verification Process*, George C Marshall Space Flight Center, Feb. 1994.

[9] NASA-STD-7002A, *Payload Test Requirements*, 10 Sept. 2004.

[10] Freeland, R., et al., *Large Inflatable Deployable Antenna Flight Experiment Results*, IAF-97-1.3.01, International Astronautical Federation, 1997.

[11] De Groh, K. K., et al., *Analyses of Hubble Space Telescope Aluminized-Teflon Insulation Retrieved After 19 Years of Space Exposure*. Astrophysics Space Science Proc. 32, 13–26 2013 DOI 10.1007/978-3-642-30229-9_2, # Springer-Verlag, Berlin Heidelberg, 2013.

[12] MIL-A-83577B (USAF), *Assemblies, Moving Mechanical, for Space and Launch Vehicles, General Specification for*, USAF Space Division, 1 Feb. 1988.

CHAPTER 3

Gravity Offloading

Gyula Greschik
TentGuild Eng. Co., Boulder, CO

3.1 INTRODUCTION

Gravity compensation (the alleviation of gravity effects) is fundamental to space structures testing on Earth: severely constrained by the cost of access to space, one is forced to create artificial conditions for a test specimen to "experience space" in the laboratory. In the overwhelming majority of cases, gravity is the most profound obstacle to making such an imitation realistic from the structural viewpoint.

Earth's intense surface gravity

- Can hide aspects of space behavior by overwhelming free play
- Can add contact forces and friction between components
- Necessitates support which constrains, stiffens, and inertially influences response
- Adds internal stresses (in the context of the support) which can alter the shape of, stiffen, or destabilize parts of the structure
- Can necessitate further compromises (for example, atmospheric effects will also be present for a test that can't be performed in vacuum)

The more compliant a component or assembly is in the context of its scale, the more detrimental the above influences can be, to the point of undermining the meaningfulness of testing (as is the case for most solar sails). In fact, by virtue of their precision requirements and/or lightweight and innovative architectural features, many modern space structures are in the twilight zone between reasonable and questionable testing solutions. For such structures which are the focus of this book, the realm of gravity compensation is a fascinating world of colorful compromises where all ideas are welcome for consideration as long as they help in any way evade, alleviate, mitigate, or at least render quantifiable the effects of

Doctor, TentGuild Eng. Co.

gravity—whether the test focuses on deployment itself, or on the precision or dynamics of a deployed configuration.

It is the writer's hope that the present review will help the reader navigate these waters. To this end, key trends are illuminated, principles to guide judgment are highlighted, and a few illustrative examples are cited. However, this summary is far from comprehensive: the points made and the examples could be easily extended. Such a limited scope (for this short chapter) is inevitable because of the rich and vivid nature of the field. Gravity offloading is ubiquitous in space structural testing: it is implicitly or explicitly addressed always and everywhere in space structures R&D, a challenge to which ever newer answers are often found. In fact, one can learn of related ideas and solutions, variations of methodology, and sometimes quantitative insight even from seemingly unrelated publications primarily addressing other focus areas. Technical sources appear endless.

Note also that this chapter provides a *general* tour of gravity compensation, rather than of its use for large lightweight applications. This approach is justified, first, by the uncertainty of what may be deemed large or lightweight (the author would personally prefer some definition involving the context of compliance). The second reason for a general treaties is that testing solutions for one kind of structure may well be worthy of consideration for other kinds, too (often, with modifications). Finally: the emphasis on "large" and "lightweight" specimens is not lost as a result of the more generic nature of this treatise because any narrative on gravity compensation naturally highlights such structures because they are the most challenging to test and offer the best case study examples.

3.2 SELECTED PERSPECTIVES ON GRAVITY COMPENSATION

Begin with a few high-level observations on what it generally takes to engineer a test imitating weightlessness.

3.2.1 THE NATURE OF THE GRAVITY COMPENSATION CHALLENGE

To compensate for the effects of gravity for a system of finite weight is to engineer the interaction of support with selected aspects of test article response. If an ideal solution were realized, this interaction would be limited to the support providing the mere uplift to precisely counter weight, without engaging the specimen in any other kind of interaction. However, perfection in this respect not only is impossible to realize but wouldn't even be desirable because it would allow the specimen to simply float away. Obviously, as a minimal imperfection a restoring bias to some initial state must be present even in offloading methods that approach (with some limitations and for some structures) compensation "perfection" with remarkable accuracy [1–3] or with a promise thereof [4].

The effects by a test apparatus on the specimen response can be seen as the combination of direct kinematic constraints, indirect kinematic constraints such as extra stiffness or energy dissipation, altered inertia, and geometric stiffness effects due to stress fields by gravity and the support. The need to mitigate these effects (combined with additional issues such as air drag, space limitations, cost, etc.) can easily end up as the primary driver of the test setup design. Due to the complexity of this set of issues and to the differences between structures and test programs, gravity compensation is typically addressed on a case-by-case basis as a pivotal, defining element of the test program.

However, despite its defining role, gravity compensation is rarely in the very focus of technical publications save when scrutinizing particular aspects of its performance [1, 3–7], often with specific solutions in mind. One interested in the state of the art in configuration and solution options is frequently forced to search for bits of information across papers on the actual structures tested.

3.2.2 DEPLOYMENT VS. DEPLOYED PERFORMANCE

In the life cycle of a spacecraft structure, two distinct time periods are "spent" practically free of gravity: space deployment and operation. Despite some overlap, rather different issues are associated with the two phases, whether performance requirements or technical-engineering challenges are concerned. These differences are naturally reflected by the verification needs; rendering deployment- and performance testing essentially two different problems. The two are routinely addressed with separate test programs.

Gravity compensation for deployment testing must accommodate transitions between dramatically different configurations. Typical objectives of related experiments may be to prove feasibility, robustness, and reliability/repeatability for the envisioned unfurling process, and to identify problems and pitfalls. (Some of the latter may actually be rendered unobservable by gravity, or they can be artificially caused by it, unless compensation is of sufficient fidelity.)

For deployment testing, response often needs to be verified only qualitatively, or only to a precision dictated by component damage tolerances.

On the other hand, a deployed structure typically has to satisfy stringent geometric tolerances and quantitatively defined mechanical performance metrics (such as vibration frequencies and tolerated system accelerations). The system configuration itself no longer changes; the structural geometry varies only as a result of perturbations. High fidelity measurements and response data acquisition are important, for which the qualitative effects of gravity as well as of the gravity offloading system need to be fully understood and tracked with precision.

It is typical to use different test programs for deployment and deployed performance testing. Despite the shared context, it is also not uncommon for hardware specimens for the two test programs to be different. The two tests may need to be coupled only if the effects of the deployment process on the final, operational, configuration are of concern. (This can be the case, for example, for high-performance structures such as optical systems.)

Accordingly, gravity compensation apparatus for deployment testing is often streamlined, first, to accommodate specimen configuration transitions with minimal constraints imposed and, only second, for quality data acquisition. This order of priorities often defines a testing philosophy associated with more "low tech" but kinematically more inventive solutions than that for testing the deployed state. In contrast, testing in the deployed state typically calls for offloading solutions to aggressively reduce and/or allow the precise quantification of the mechanical influences by the test apparatus, while tracking transient or quasi-static structural responses with high resolution and detail.

3.2.3 ASSESSING AND QUANTIFYING GRAVITY AND TEST EFFECTS

The mechanical influence on specimen response of the gravity field and support loads, combined with the inertial, constraint, and dissipation effects of the test apparatus, can be readily quantified with standard mechanical modeling abstractions. After all, both the assembly of interest in space and its version combined with a support rig in gravity are structures. Mature methods of characterization are available for both.

Geometric stiffness due to gravity-induced stress fields in various components and sub-assemblies can be combined in standard numerical models as matrix quantities with inherent structural stiffness and that from design preload, if any. As such, for any test setup, these effects can be easily related mathematically to the specimen's own stiffness in the context of whatever degrees of freedom one desires to use. The formulation of support stiffness effects (the forces and moments with which the apparatus responds to displacements, divided by the latter) is similarly straightforward.

Akin to stiffness, inertia and damping by the support can also be easily integrated into the mathematical model, yielding standard and less than standard formulations in a straightforward manner [6, 7].

Consequently, some characteristics of the structural model combined with apparatus effects derive easily from basic principles of mechanics. For example, of stiffness, damping, and inertia directly added to the specimen, only the first, stiffness, can be negative if the support system is passive (i.e., if no element of the apparatus is actively powered or actuated). For this case, model inertia and damping can be increased only by coupling the specimen to the apparatus. For a support system with active elements (where actuators either produce or "tune" the supporting forces), apparent negative effective values for all three (stiffness, mass, and dissipation) effects are possible.

The ratio of the inertia (of the support system effective mass) added to that of the specimen mass can often be reduced to below 5 percent [1, 3], or even less [4]. However, this fraction can steeply increase with a decrease of specimen mass.

Second order stiffness effects can also be added to the test setup by the suspension system. A typical case is suspension from a fixed location of a point on the specimen that laterally moves. In this case, a "swing" stiffness will emerge, creating a restoring force in response to any horizontal movement of the suspended anchor location.

For vibration testing, the easiest way to eliminate such geometric second order swing stiffness is to orient the test article so as to render the anticipated vibration response vertical. (Of course, the suspension in such a case has to be compliant, which can be achieved by integrating a soft spring, or some other device, in the suspending cable.) A useful metric of apparatus performance in such cases is how the vertical plunge frequency of the supported specimen relates to that of the structure in true weightlessness. Kienholz [1] recommends that the suspension frequency generally be at least an order of magnitude below the test article's first flexural natural frequency. He also notes that, generally speaking, specimen frequencies above 20 Hz don't call for anything more sophisticated than suspension by simple springs. However, such a solution for specimen frequencies less than 10 Hz or so can lead to static sag and spring surge modes. In addition, the support problem becomes increasingly difficult when frequencies an order of magnitude less need to be accommodated. Some of the lightweight, compliant structures characteristic of current aggressively innovative designs belong to the last realm.

Kienholz's work [1] also offers an excellent summary of performance considerations and practical limitations, beyond the narrower scope of stiffness, mass, and damping influences. Its systematic review of practical effects in the context of the quantitative objectives is an invaluable reference.

Having acknowledged the maturity of the mathematics of and the theories for the formal treatment of gravity and support effects, one must also note that the numerics for such problems is often less than straightforward. The reason behind this discrepancy is that when theory is translated into "number crunching" to solve a problem, numerical stability can be undermined by mechanical or numerical ill-conditioning. Calculations can also be made very expensive by nonlinear effects beyond basic geometric stiffness; these nonlinearities include nonstandard dissipation mechanisms and contact phenomena. Numerical and mechanical ill-conditioning as well as complex non-linear phenomena are characteristic of coupled compliant model and test apparatus systems. As a result, the reliable quantitative investigation of gravity and apparatus effects can be a major exercise by itself. However, the advancement of computer simulation tools keep pushing onwards the boundaries of affordable—or, at least, manageable—numerical analysis. Even for complex structural systems, it happens with increasing frequency that numerical analysis reveals ahead of time some of the pitfalls that would later be exposed by hardware testing. (For example, this happened during the Power Sail project [8].)

3.3 GOLDEN RULES

Even if well versed in the art, one may benefit from occasionally rising from the details and contemplating principles that, on some level, generally guide creative engineering work. For the less than fully initiated, a high level perspective is even more useful because it can shape one's view and understanding. The following set of "golden rules" has been compiled in this spirit to capture fundamental elements of good practice in the field.

3.3.1 DON'T TRIVIALIZE THE CHALLENGE!

In most projects, it is tempting to pay the minimal possible attention to engineering a gravity compensation system. This is understandable because, in the ultimate context of the space mission served, the design, fabrication, and operation of a test apparatus are of a mere infrastructural element with utility limited to some phases of development. They are only remotely related to the actual future mission objectives. The view of testing details as of secondary importance is also reinforced and reflected by the focus of most publications produced on the project, once complete. Public interest naturally turns toward the final system, leaving development technicalities to specialized technical publications, if any.

In fact, under the pressure of aggressive schedules and overwhelming workloads, the developer also, in the early phases of work, may be glad to simply see some (or any) proof for the feasibility of a concept. At such times, rudimentary "quick-and-dirty" testing may indeed be sufficient.

However, such an approach can rarely suffice for reliable and final system or component verification. For such tests, the design, verification, and optimization of the gravity offloading system itself become key program elements, both enabling and quantitatively critical to the success of the project; their importance cannot be overlooked without risking project success. Finances and attention sufficient for the quality engineering of the gravity offloading system should be allocated to the task.

3.3.2 UNDERSTAND THE STRUCTURE!

A precondition of successful engineering—in fact, of any endeavor—is the proper understanding of the subject. For gravity offloading system design, this includes an in-depth understanding of the test article structural mechanics.

One may distinguish two aspects of this understanding. The first one is insight into the expected response via engineering judgment (experience) and the proper interpretation of kinematic, mechanical, and structural models. From this insight, a realistic anticipation of general specimen and apparatus behavior results.

For example, one may foresee the deployment kinematics, the coupled nature of vibration modes between the whole specimen and some of its compliant components, or a possible interaction of torsional and flexural modes of some long members. From this understanding, one can develop a general idea of an acceptable framework within which support solutions shall be sought.

The other kind of understanding called for when engineering a test setup is an understanding of the sources of trouble: of the sources of possible or likely mismatch between expectations and reality. These sources include the limitations of the analytic models used, and the kinds of specimen and apparatus failure modes that may occur. An intelligent awareness of the risk of unforeseen failure events is also necessary. The development of escape strategies, and the selection of a data acquisition system able to capture valuable data in the case of unforeseen events, can only be based on such insight.

The recognition of the need for multiple tests that target different aspects of specimen response (including response associated with different specimen orientations) is typically also based on the understanding of the involved mechanics.

3.3.3 UNDERSTAND THE ENVIRONMENT AND THE REQUIREMENTS!

To faithfully represent key events and to cut corners only where compromises are permissible, it is critical to have a system-level understanding of the environment and the disturbances imposed on the hardware in the simulated condition, and to understand the aspects of performance that critically influence the mission performance itself. For example, for the testing of elastically actuated undamped deployment (where transient kinematics may invite hardware safety issues), suspension must be aggressively streamlined to minimize added inertial, stiffness, and energy dissipation effects. On the other hand, when deployment is slow and controlled, the inertial influence by a support system is less of a concern, and quasi-static motion may be acceptable.

3.3.4 APPROACH THE PROBLEM INCREMENTALLY!

A direct derivative of the appreciation of the challenges of test apparatus design is that its engineering follow professional standards and protocol. Such a formal approach often requires the initial concept verification of the apparatus idea itself. One may first construct a simplified, reduced-scale, support and observe its bread-board performance with simplified mock-up specimen hardware. Such initial efforts, although rarely published, offer payoff as significant as that of the final testing itself. Concept flaws, if any, are revealed early and confidence in the support architecture is gained before the invested effort irreversibly couples the design to any specific paradigm or detail.

Sometimes, incremental testing also includes orbital deployment tests. A good example is the 13 m aperture diameter mesh antenna deployed in 2006 [9] on board the Japanese Engineering Test Satellite VIII (ETS-VIII), with a support structure of adjoint hexagonal truss modules. Despite a thorough, systematic, and incremental ground development and testing program, the first orbital test, in 2000, still ended with the structure locked in a partially deployed state. The forensic study concluded [9] that the most likely source of the problem was the limited recognition of how gravity and the gravity compensation setup constrained during ground testing some transient response elements. The latter ended up more significant in space than expected, likely causing one or more cables to get caught. A second orbital test successfully deployed later, and the final hardware performed well. See Chapter 7 for further discussion on this example.

3.3.5 VERIFY THE TEST SETUP WITH ANALYSIS!

Surrogate testing (hardware testing to verify analysis methodology) has become a well-established practice in recent decades, supporting the replacement of flight hardware testing with simulation (primarily, FEA [finite element analysis]) when the former is not affordable or feasible (often due to gravity issues).

However, less recognized is the help that structural analysis can offer in the very design of a test and that of gravity offloading setup. This is unfortunate because test apparatus design can be as challenging as the design of the tested hardware itself (although the risks and costs are different). Accordingly, quality engineering attention, complete with numerical simulation, can highly benefit apparatus design by illuminating specimen-equipment interaction issues that may otherwise be unnoticed until testing begins. This can enable apparatus optimization and can predict test fidelity. Such analyses also help one to understand specimen response to the gravity in the context of a support scenario. The costs of the analytical design of a test setup are naturally reduced by coupling the effort to the analyses of mission deployment or performance, by "recycling" elements of the specimen numerical model.

Beyond providing insight, another, critical, role of the high-fidelity analysis of a test setup may be to spot and thus prevent strength (stability) failures in equipment and structure. (Lightweight space hardware must be supported without the gravity loads between support points crippling the hardware. In addition, the elements of an offloading apparatus streamlined for minimum weight [4] must also withstand gravity loads by the tested structure and themselves.)

McEachen and Trautt [10] present an example for the mutual verification of 40 m long lattice boom flight hardware and its analysis in a test environment. The numerical model for the horizontally suspended specimen includes the air bearings that allow lateral horizontal movement, and auxiliary structures of suspension from the bearings to the mast. In order to establish trust in the analysis, systematic testing was used to calibrate some aspects of the numerical model.

3.3.6 USE THE RIGHT FRAME OF REFERENCE!

The trivial fact that there is no up or down in space has a more general analogy that can be easily overlooked. In true weightlessness no particular part of the structure may be associated with the inertial frame by default. Consequently, representing the spacecraft (or bus) mount of a deployable structure in the laboratory with a fixed mount (one on a wall or a rig) can severely distort (deploying or deployed) specimen dynamics, regardless of the design and quality of the gravity offloading apparatus. For a faithful representation of behavior in space, the system of interconnected spacecraft parts should be suspended in its entirety, with no part fixed.

However, even in tests to validate only kinematics (that need not reflect the proper system dynamics), the choice for the testing frame of reference should not be taken for granted. It is possible that one reference frame can eliminate difficulties associated with others. For example, to test the unfurling of an appendage of a certain length and construction, rather than holding either end steady, it may be simpler to orient the assembly in space by holding firm an intermediate location from which both the bus and the outer end will rotate away.

The frame of reference used should be considered a variable of test engineering design, rather than a predetermined parameter.

3.3.7 NOTE "HIDDEN" GRAVITY EFFECTS ON THE SPECIMEN!

When being kept from falling, a test article experiences more than the external kinematic constraints and/or parasitic (inertia, stiffness, and dissipation) effects directly imposed by the supporting equipment. As already alluded to above, internal response to the pattern of gravity and supports can beget second order effects. These effects manifest in a model as (positive or negative) stiffness added at various internal or external degrees of freedom, whether at locations of articulation or elsewhere. In some cases, these effects can be seriously aggravated by the combination of test setup and specimen orientation.

For example, in a specimen vertically suspended for testing, tension by gravity accumulates upward. The accumulated loads, near the suspension at the top, can, in cases, be high enough to seriously influence specimen response, especially for thin, lightweight structures such as sheets. In such scenarios, workarounds have to be explored. For example, interim support at locations along the suspended length may be introduced.

A specific example is the first orbital test of the mesh antenna for the Japanese ETS-VIII. The residual mechanical influence of gravity (specifically, on the unsupported cable network) in the laboratory to which the failure to identify in the laboratory the deployment problem encountered in space [9] was attributed is a form of geometric nonlinearity: a "hidden" gravity effect.

3.3.8 DEVELOP HARDWARE THAT CAN BE TESTED

A hallmark of good engineering is to create solutions that lend themselves not only to the service they are intended to provide, but also to their own physical realization so they can be easily built. The system has to be envisioned with its development path in mind. With the critical role of gravity compensation in the testing of large compliant space structures, this principle implies that the structural concepts and solutions should be conceived with their future testing program and offloading needs in mind.

In some cases, architectural decisions have also been made with explicitly considering this principle. For example, the reflector module sizes for the antennas [9, 11] on board the Japanese ETS-VIII satellite were defined to facilitate testing.

Unfortunately, the more ethereal a structure, the more severe the limitations to how closely this guideline can be followed. In extreme cases, easy testing is out of the question. Design and engineering in such scenarios can, at best, be streamlined toward the proper management of foreseeable compromises.

3.3.9 BE READY TO COMPROMISE, AND DO IT CREATIVELY

It is a rare luxury in testing programs to have the resources to build apparatus and execute tests that can properly answer all questions of interest, even if such tests are technically possible. To still get the best out of testing, a keen awareness of the conflict between demand and resources is required, and to achieve the best results, one must have an open mind to navigate the constrained design space.

Often, even the duality of deployment and deployed performance testing cannot be accommodated; despite the dramatically different needs of testing for these two phases of performance, one may be forced to use a single test setup. One answer to this challenge may be some form of modularity: within the same setup, one set of support elements for deployment testing and another for mission performance verification may be activated. For example, deployment may be tested with a suspension system with a minimal number of movable support locations. However, refined or alternative support (such as compliant springs at optimized locations) may be employed for higher fidelity when the vibration of the deployed system is tested.

The readiness to compromise also has to extend to the quality of both test and analytical results. In some cases, even a 40 percent agreement between predictions and physical measurements has to be considered fair [9]. A keen eye, knowledge, engineering intelligence, and the ability to discern critical trends from secondary details are needed to rely with confidence on less than ideal results.

An example for smart test design for limited circumstances is the systematic testing of a 40 m coilable lattice boom for load combinations as well as dynamics [10]. For this test, it was not possible to provide a sufficiently sensitive two-axis gravity offloading system permitting free lateral and vertical motion. As a solution, the boom was positioned horizontally and was supported with air bearings, allowing only horizontal motion. The specimen was rotated for testing in multiple orientations, and the test results were then combined.

3.3.10 UNDERSTAND THE TEST AND THE TEST EQUIPMENT

To avoid (or enable the control of) conflicts between the limitations and capabilities of devices and components, it is of cardinal importance to understand the apparatus and support solutions. As an example, consider the vibration testing of a large, horizontally suspended appendage. To gauge the passive response after a step-like disturbance, suspension via compliant springs (or "bungee cords") may be appropriate. However, when excitation patterns are considered, a more advanced scheme may be necessary because the suspension cords themselves might respond to the excitation with resonance at certain frequencies, foiling the test.

The specific example of a mismatch between test results and analytical predictions for L'Garde solar sail development hardware in the early 2000s is also worth recalling. When the problem proved nontrivial to resolve, the writer (who had earlier authored the study [12] enabling the development of that architecture and subsequently invented the sail mechanical structure itself [13]) was approached for advice. The cause of the mismatch was identified as a seemingly minor approximation in the FE modeling of the suspension system fly beams. Namely, the latter had been modeled idealized, ignoring the cord attachment point offsets from the fly beam centerline. (For more information, see the "Fly Beams" section, later in this chapter.) After the adjustment of the numerical model, acceptable agreement with test results was achieved [14].

3.3.11 DON'T BE SHY TO SEEK ADVICE

Finally, note that advice—even informal comments—can call attention to important issues that the test engineer may overlook when being submerged in details. Informal peer review can usually greatly complement one's expertise; it should be sought even if the need for it is not critical.

3.4 A REVIEW OF COMMON GRAVITY OFFLOADING MEASURES

One could classify gravity offloading approaches in different ways. In fact, the ideas offered over the decades, as well as their advantages and shortcomings, are so diverse that one may wonder how much sense it might make (beyond the "impromptu" didactic streamlining of a discussion) to propose a structured view of this field at all. For the sake of the present summary, one objective of which is to illustrate the diversity of concepts, a couple of approaches to the classification itself of offloading solutions will here be reviewed.

Examples of actual offloading measures will be employed to illustrate some the ideas and options mentioned. Because most of these examples could well be cited in the context of more than one of the points made, some of them will indeed be cited multiple times. Others, on the other hand, will appear only once, even if they have aspects that may fit the narrative elsewhere also.

3.4.1 THE DIMENSIONALITY OF SPECIMEN MOTION

A simple way to classify gravity compensation measures would be based on the dimensionality of the specimen motion accommodated [15]. Accordingly, one can distinguish support systems to permit linear, planar, or spatial kinematics. (For a discussion on structural classes, including dimensional classifications, see Chapter 2.)

3.4.2 THE TYPE OF MEANS USED

An alternative, similarly broad, classification could be based on the mechanical principles on which offloading schemes are based. For example, one could distinguish [5, 16] physical, buoyancy, pneumatic, and mechanical methods. (The first category, "physical approach," includes parabolic flight and drop tests. This is not literally gravity compensation but, in fact, orbital tests with Earth-colliding orbits.) In the following subsections, means of compensation are discussed in an order very coarsely aligned from the least to the most purely "mechanical."

3.4.2.1 FREE FALL: DROP TESTS

The state of free fall is, by definition, weightlessness: being in orbit is free fall with an initial condition that causes one to continuously miss the Earth (the knack for flying "lies [in] learning how to throw yourself at the ground and miss" [17]). However, on Earth, the duration of free fall from a height h is $t = 2h/g$, with g being the acceleration of gravity—less than $t = 1.5$ sec for $h = 10$ m height and about $t = 4.5$ sec for $h = 100$ m. Although this net fall time can be doubled by launching the test upward, the time bracket and the need to manage specimen launch and catch are severe limitations. Accordingly, drop tests in a laboratory environment are rare. Even the world's tallest vacuum-capable drop chamber, the vertical shaft for a maximum of 132 m free fall in 5.18 sec (the Zero Gravity Research Facility at NASA Glenn Research Center [18]) is generally used for the study not of mechanical phenomena but of the effects of microgravity on physical and material combustion, fluid physics, biotechnology, and so on.

An example for structural drop testing is that of the elastic deployment in a 1.3 sec fall (from 8.2 m) of a truss unit in 1980 [19]. Although the specimen was not scaled down (the truss consisted of 36 struts each a little more than 2.1 m long, packaged in a bundle of 4.3 m length), the deployment was fast enough to complete in free fall. One observation was that reducing the air pressure to $1/10$ atmosphere already rendered air drag effects on the deployment negligible (harder vacuum would be necessary for more compliant and lower weight structures).

High-altitude drop tests and sub-orbital flight parabolic trajectories are special cases of drop test; the latter may still be considered a laboratory environment, on board an airplane [20]. Beginning with the late 1950s, aircraft in various countries have been used increasingly often to produce zero- or reduced gravity environment for testing and entertainment. By the 1960s, the periods of weightlessness routinely achievable reached 25 to 30 sec.

Reference [21] describes an early deployment experiment of a deployable lightweight film solar array structure and contains interesting historic information on this flight technology in the United States.

3.4.2.2 BUOYANCY IN AIR: BALLOONS

Balloons, a simple way to make something float, have some impressive features. They inherently impose no kinematic constraints, add no stiffness, and are cheap. Balloon support is a great "poor man's" offloading system, eminently applicable to some static or quasi-static problems.

The two most obvious limitations of the method are spatial constraints and air drag. The former limits application to where there is room for the balloons (where balloon size conflicts neither with the spacing of specimen support locations, nor with room size). The latter limitation restricts the method to problems where the tested structure only slowly moves.

When moved through air, a balloon doesn't experience drag only in the sense of damping; rather, its effective inertia is also increased with an extra apparent mass due to the inertial effects of the air that is forced to flow around it. In a standard room environment, the balloon mass itself (with the gas within) amounts to a direct mass overhead that cannot be less than 16 percent of the payload mass [1, 22]). The inertial effect of the air flowing around the balloon (an effect long known in ocean engineering in the context of the forced movement of submerged bodies [23]) is equivalent to further adding to the system about 77 percent of the offloaded mass [3].

Another, less unique, drawback of balloon suspension is the geometric stiffness effect, when the balloon follows the movement of the anchored location with an off-vertical suspension cord angle. This effect amounts to an incipient lateral stiffness that combines with the drag and inertial overheads and renders the influence of suspension on the test dynamics different in the vertical and horizontal directions.

Yet another disadvantage of balloon suspension is atmospheric effects. Air movement must be minimized, and testing calls for frequent checks and tuning, possibly multiple times a day, because meteorological conditions (atmospheric pressure) influence balloon lift. However, even if atmospheric conditions are steady, balloons require daily adjustments to maintain a desired lift force because of helium leakage.

The balloon gravity compensation for the compression testing of a slender deorbiting sailcraft boom [24] at the University of Surrey, UK, is shown in Fig. 3.1. The solar sail mast of a composite "carpenter tape" cross section is one of four, unreeled from the hub in the foreground, with its far end loaded as shown in Fig. 3.2. (One of the other masts, deployed downward, is also seen in its entirety, folded near its base and laid on the table; only the bases of two masts are visible.) Note the cups under the balloons, for weights to tune the lift.

3.4.2.3 BUOYANCY IN WATER

Although the buoyant effects associated with full submersion in water are mostly used to offload testing manned space activity [25] (extra-vehicular activity = EVA),

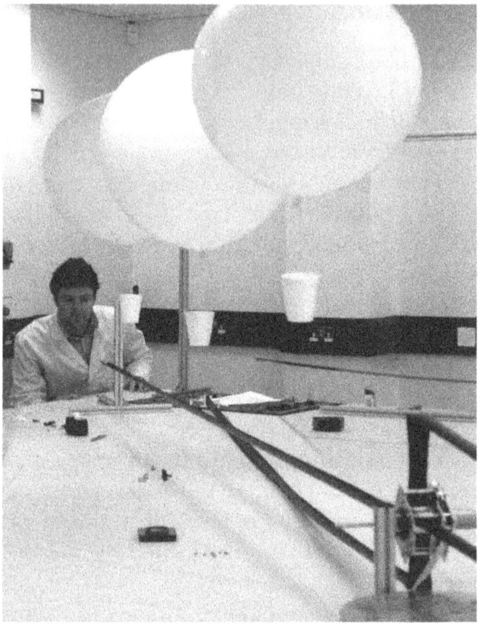

Fig. 3.1 Solar sail composite tape boom buckling test with balloon support. (Courtesy of Gabriel Secheli and Andrew Viquerat, Surrey Space Centre, University of Surrey, UK.)

the water *surface* offers excellent support if only one- or two-dimensional response is to be investigated, and the added inertial, damping, and stiffness effects can be ignored. These limitations, however, are quite severe given water's density, the drag it can exert on rapidly moving bodies, the buoyant stiffness in the vertical direction, and possible wave dynamics in response to specimen behavior. Accordingly, water support is practically limited to certain quasi-static deployment tests.

An example for such an application is the deployment verification of L'Garde's conical inflatable boom technology for a solar sail mast [26]. This boom was an order of magnitude longer than the scale for which its technology had been verified earlier [27]; the test was performed in a water trough.

Fig. 3.2 Loaded far end of tested boom that appears in Fig. 3.1. (Courtesy of Gabriel Secheli and Andrew Viquerat, Surrey Space Centre, University of Surrey, UK.)

3.4.2.4 PNEUMATIC SYSTEMS

Pneumatic and hydraulic systems are often used in power transmission, as well as in force amplifier and force attenuation equipment. A key difference between the two is the inherently elastic response of the former, by virtue of the compressibility of gases. In fact, for a given system volume change (such as a given piston stroke), the associated pneumatic "spring coefficient" can be arbitrarily reduced by simply increasing the system volume (e.g., by linking with pipes additional gas containers to those already used). Therefore, because the reduction of support stiffness is one of the objectives of the gravity offloading problem, pneumatic support is an attractive gravity compensation option.

Beyond simply casting a suspension cable over an inflated "bag," the pneumatic principle is typically used in offloading via a pneumatic-magnetic ("pneu-mag") device, sometimes informally also referred to as the "Kienholz machine." This device was initially developed by CSA Engineering (now Moog-CSA) in the late 1980s and early 1990s [1, 28, 29] for the structural development of the International Space Station. The device has since become a standard piece of high-technology equipment in space structures testing. Of each of its load-bearing units, the primary component is a pneumatic piston (or pair of pistons) that, coupled to an external tank, bears the allocated test article weight with very little added stiffness. In fact, even this stiffness is reduced to a small perturbation by regulating pneumatic pressure—adding more air or letting some escape, to compensate for even minor pressure increases or decreases that may result from the motion of the piston (i.e., of the supported specimen). Piston friction is eliminated by air bearing (a leak between piston and cylinder). Thus the supported weight, regardless of its magnitude, is upheld with minimal dynamic stiffness and virtually zero friction. This core pneumatic system is complemented by an active electromagnetic subsystem that provides gentle centering control and compensates for minor perturbations and imperfections associated with the pneumatics. In a pneu-mag device, two different subsystems are combined, each doing what it does best.

A simplified pneu-mag device load-bearing unit is shown in Fig. 3.3 via an illustration of its structure on the left and a front view photograph on the right. The pneumatic piston and cylinder are at the center of a vertically movable rectangular frame called the *carriage*, which is coupled with the load cable to the test article below. The unit shown is simplified: instead of the active electronic subsystem, it contains only a centering spring, sufficient for many tests and less costly. Not shown are the rest of the pneumatic system (tubes, tank, and control components).

The active system itself consists of position, velocity, and acceleration sensors; control electronics; and an electromagnetic (voice-coil) actuator. This subsystem can effectively reduce frequency-dependent stiffness effects and mass cancellation. It can also provide measured, controlled disturbance input for high-fidelity dynamic testing.

Several device versions [1] have been developed from the concept and are applicable to various conditions. Although the initial application of the

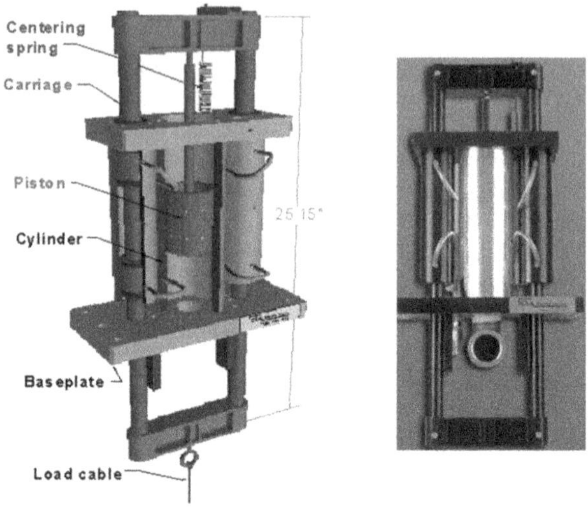

Fig. 3.3 Structure and front view of simplified suspension unit of pneu-mag offloading equipment, wherein a passive centering spring is used to provide some of the functionality of active electromagnetic control. (Moog-CSA Engineering Model AGM-350-6 [30].)

pneu-mag technology has been vibration testing that involves small displacement magnitudes, devices with large stroke can also be fabricated and have been applied to large motion tests such as the ground-based testing of the tracking and pointing experiment for a 7000 lbf truss testbed structure [1].

3.4.2.5 MECHANICAL SYSTEMS

Mechanical apparatus is by far the most common offloading option; it offers diverse solutions for many common space structural test problems. Accordingly, it is not discussed here simply as a sub-case of the physical means possibly employed in gravity compensation. Rather, various aspects of mechanical systems and mechanical elements are covered in other sections and subsections in this chapter.

3.4.3 LEVELS OF CONFIGURATION COMPLEXITY

Yet another way to categorize gravity offloading options may be based on the complexity of specimen arrangement and specimen motion. Accordingly, one can distinguish the following eight approaches.

3.4.3.1 SCALING

The simplest route to offloading is to avoid offloading at all by downscaling until, according to constant thickness [31, 32] or classic proportional [33] scaling principles, gravity effects reduce to tolerable levels for the specimen with reduced size. This approach is the most attractive if sensitivity to minor perturbations and quantitative fidelity are not critical, and if the system to test is easy to scale. This is the case for many membrane structure deployment tests, especially those employing centrifugal force, because the spin rate can be easily increased according to the reduction of system dimensions.

Two examples for tests where gravity effects were eliminated with downscaling are a 2001 study of configuration variations for spin deployed segmented membranes [34] and the 2006 investigation of some dynamic characteristics of the spin-deployment of circular membrane sheets packaged with the rotationally skew fold [35]. Both served conceptual solar sail development and were performed with scaling for convenience, without the need to quantitatively match precise deployment characteristics of full-size space hardware.

3.4.3.2 DROP TESTS

Another principally simple solution is to avoid gravity effects altogether via a drop test of some sort. However, a claim of low complexity may be misleading in this case. Although "offloading" is simple for a drop test, other issues challenge the test design. These include test time (drop time) limitations, the setup effort before each test, and the difficulties of instrumentation. Drop tests were discussed in the "Free Fall: Drop Tests" section, earlier in this chapter.

3.4.3.3 VERTICAL ALIGNMENT

The next level of complexity above avoiding gravity, applicable to some specimens of longitudinal or planar shape, is simple vertical alignment (in the direction of the gravity field), with the specimen simply hung in this orientation from its upper end or edge (although cantilever support by the lower end is also theoretically an option). The support itself can be chosen for the convenience of the test. It may even allow rotation and horizontal motion, such as via a hinge on a "cart."

Such support poses no (or little) constraint in the lateral direction, and the specimen alignment eliminates the direct gravitational bending of the specimen. This simple setup may be acceptable if the geometric nonlinear effects invited by the alignment are acceptable—typically, if the specimen's own flexural stiffness is high enough to render its effective alteration by the gravity field acceptable.

The nonlinear effects invited by vertical suspension (or cantilever support from below) result from tension (or compression) building up in the specimen. These effects add to the structure's geometric (second order) stiffness against lateral displacements. This change is positive for suspension, stiffening the lateral response with restoring "pendulum stiffness." Lateral vibration frequencies will be driven higher than would be without gravity. For cantilever support from below, the added stiffness is negative, and lateral response is softened.

Fig. 3.4 Test setup for deployment upward, with pulleys-and-counterweight offloading. (Courtesy of Gabriel Secheli and Andrew Viquerat, Surrey Space Centre, University of Surrey, UK.)

The effects of gravity on a deployment test in the vertical direction are direct, assisting (accelerating) downward deployment and hindering upward motion. For an end-supported vertically aligned specimen, this can be acceptable only if the deployment dynamics comfortably overwhelms gravity effects; this may be the case for small elastically self-deployable structures, such as some CubeSat applications. Otherwise, simple offloading can be added for upright deployment, such as shown in Fig. 3.4 for the inflatable deployment test of a boom.

In fact, for the test in the photo, deployment happened too fast for the counter-mass to exert its full weight on the cord while dropping. Consequently, spring-based low inertia offloading was contemplated for subsequent testing [36]. The observed force reduction due to acceleration well exemplifies the need for test setups, offloading included, to be considered as single, integrated systems of specimen and apparatus—even for simple configurations such as that shown in Fig. 3.4.

A more complex schematic for a vertical deployment test, with multiple specimen offloading locations, is shown in Fig. 3.5 for a test performed in the 1960s. Such arrangements are rare; vertical alignment is most often used for simple setups.

3.4.3.4 HORIZONTAL ALIGNMENT

If a vertical test configuration isn't practical or acceptable, the route commonly followed is horizontal alignment with distributed support.

Horizontal orientation frees the specimen from direct geometric nonlinear effects on most or all of its members, and horizontally "spreading" the structure offers comfortable configuration options for the support. This is by far the most common approach to medium- to high-quality testing, associated with seemingly endless variations and solution ideas. In part again guided by a perception of increasing apparatus complexity, one may choose to highlight the following option elements to exemplify some recurring themes.

Stiff Suspension from Fixed Locations

Simply hanging the tested structure from above with stiff tension lines constrains the specimen anchor points in the (typically, vertical) suspension direction without suppressing lateral motion. Some stiffness ("pendulum stiffness") against lateral displacements is added, via the same geometric nonlinear mechanism that emerges in the specimen's own context in typical vertical test arrangements.

An example of such support application, shown in Fig. 3.6, is a membrane optics structural model for the testing of vibration control and shape control and its analytical modeling [38]. The primary aperture structure, on the left, is

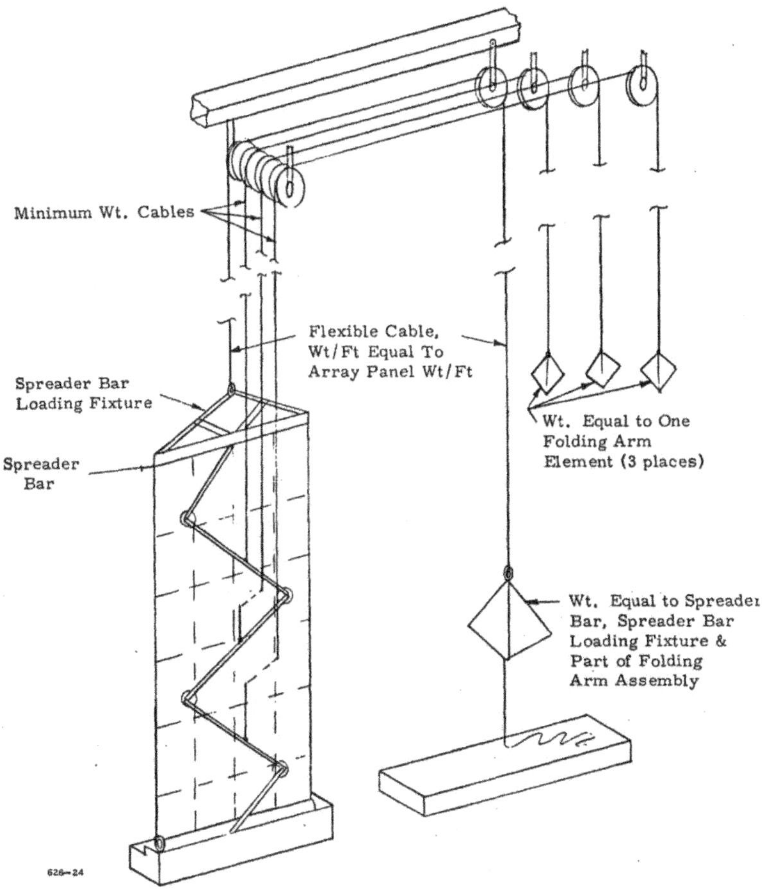

Fig. 3.5 Vertical deployment test schematic with multiple suspension points [37, Fig. 2.5.2-1].

Fig. 3.6 Vertical suspension with rods (on the two sides of the torus stretching the membrane aperture) and with cable (in the focal plane) of membrane optics prototype with inflatable support [38, Fig. 5].

a circular membrane sheet stretched within a $D = 3.8$ m diameter inflated perimeter "torus." The place of a (hypothetical) sensing platform is on the right, with a hexapod structure of inflated struts connecting the two.

In fact, this structure is a characteristically spatial one for which the nomenclature for horizontal or vertical orientation is not *a priori* unambiguous. This test is herein referenced to exemplify horizontal alignment because of the support is "spread out" horizontally: two rods at diagonally opposite points of the pressurized ring structure, and a cable at the sensing platform. (From the viewpoint of the primary aperture alone, this arrangement should be identified as vertical.)

Although the cited publication [39] does not discuss whether the support cables are compliant or actively controlled, secondary information suggests stiff support. Namely, for the final test setup (Fig. 3.6), the two cables on the sides of the aperture were replaced with steel rods; if support compliance had been an objective, this wouldn't have happened.

Vertically Stiff Support Underneath

If specimen support locations can be selected such that their displacements are (by the nature of their kinematics) restricted to a given plane, it can be an attractive solution to let these points horizontally move with no or minimal friction. Support can be direct (if the hardware support locations are directly "placed" on some surface underneath) or indirect (when fixtures sliding on a surface hold the anchor points).

A typical means of such support is an air table or air bearings, where pressurized air is continuously pumped through small holes on the surface to form a thin near zero-friction layer on which the hardware can easily slide. This method is widely used for spacecraft and robotics control simulations in general [40], but applications are typical in vibration and deployment testing also [10, 41, 42]. Figure 3.7 shows a snapshot of the deployment test of an inflatable beam on an air table.

Although several precision high-technology air tables are available on the market to assist such tests, the fundamental principle is so simple that even low-cost, "homemade" equipment is often sufficient. An example is a dynamic test program of spin-deployed membrane disks [43] where a 1×1 m air table was constructed to support the specimens during spin-up. Once the desired rotational speed was achieved, the table was lowered from its initial position under the spinning membrane, allowing testing to proceed unhindered.

An analogue to air table testing is the elimination of support friction via dither: via a fine but intense vibration of the surface to eliminate friction effects [44].

Direct sliding support, with no mechanical or pneumatic friction reduction, can also be attractive due to its simplicity. In part, sliding support was relied upon during elements of the testing of the $D = 5.7$ m diameter Flexible Unfurlable Refurlable Lightweight (FURL) solar sail, designed for repeated autonomous deployment and retraction. One key design objective for this structure, ground testing without offloading equipment [45], was achieved in the hardware orientation with the four sail booms above the sail sheet. However, due to the asymmetry of the TRAC boom cross sections used, support was needed in the "flip

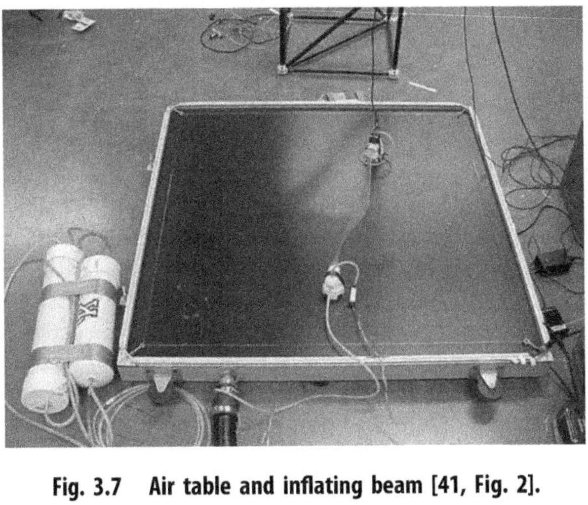

Fig. 3.7 Air table and inflating beam [41, Fig. 2].

Fig. 3.8 Sliding surface under solar sail test hardware [46, Fig. 1].

orientation," when the booms were under the sail sheet. This was achieved with a smooth sliding surface [46] (the white sheet under the sail structure in Fig. 3.8) that supported the deploying booms by keeping them in the deployment plane.

This test also well exemplifies that, for a horizontal sliding surface to be applicable, not all elements of the hardware need to move strictly horizontally; only the supported locations should be so restricted. For FURL, the membrane itself deployed with a kinematics with critical spatial aspects in the details. Only the ribs, not the deploying membrane, were supported from underneath: the kinematics of the latter were planar.

Support via Buoyancy

Buoyant support was discussed previously in this chapter. It is mentioned here only for the completeness of the list of common horizontal specimen alignment options.

Compliant Suspension from Fixed Locations

Another simple and attractive solution to gravity compensation is to suspend the specimen from fixed locations, but in a compliant manner: typically, with soft springs of elastic cables ("bungee" cords). An example for this method is the testing of the vibration characteristics of a box beam, performed as a step of a wide and thorough test and analysis project of the PowerSail deployable solar array mechanism in the early 2000s [8]. To minimize support effects, the beam was suspended at the expected nodal locations of its first, dominant, free-free vibration mode, as shown in Fig. 3.9.

For linear suspension from fixed locations, the same second order "pendulum" stiffness effect arises, in the lateral direction, as is characteristic of support with stiff cables. However, the replacement of the "rigid" direct support of a stiff suspension cable with a compliant suspension line doesn't entirely eliminate the support stiffness in the axial direction. Rather, the support stiffness is only reduced.

The simplest way to reduce both pendulum stiffness and axial residual stiffness is increasing suspension line length. The longer the spring/cord (typically, the greater the suspension height) with respect to the range of specimen motion, the "more constant" the support reaction and the less significant the pendulum stiffness. However, suspension height reduces the cords' natural frequencies and, accordingly, renders the system more sensitive to transient perturbations.

There are no general rules of applicability; options have to be considered in the context of the test. A nice example of a very simple yet effective setup is the 2010 test of the transient deployment of array panels from the sides of a CubeSat [47]. A long, soft spring via a cord suspended each panel, from a single support point 14 m above. In fact, soft spring suspension for this test was selected for its simplicity. Strings were attached to the top edge of each panel above the centroid. A very soft spring was placed in line with each string, pretensioned (stretched) to provide the right offloading force. Due to very high placement of the external suspension point, the cords needed only to offset from the ideal vertical direction with 2.1 degrees. The polluting acceleration associated with this offset was assessed and found to be much lower than the acceleration of the deploying hardware; thus, the geometric nonlinear effect was deemed acceptable.

Balancing Specimen Weight

If the inertial pollution of a test isn't of prime concern (or, at least, it can be tolerated when precisely known) and if transient acceleration in the support direction doesn't exceed that of gravity, simple counterweights offer a practicable offloading solution. By this method, effectively zero support stiffness in the direction of suspension can be achieved.

Fig. 3.9 Beam suspension at expected vibration mode nodal locations [8, Fig. 6].

Fig. 3.10 Deployment of solar array from Z-folded package on CubeSat side, with the panel contours shown in white [47, Fig. 14].

The counterweights exert lift on the specimen via pulleys or fly beams. The former doesn't inherently limit the range of vertical motion but may add friction; the latter is friction free but restricts motion by fly beam length because if a fly beam becomes too steep, its operability is jeopardized. Aside from the difference

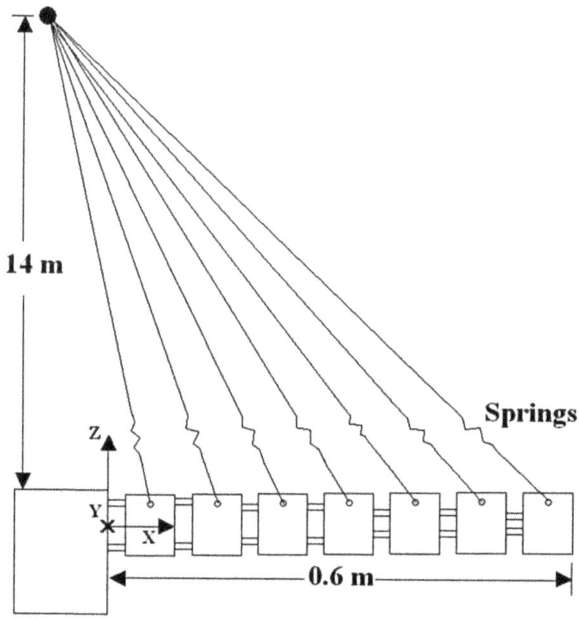

Fig. 3.11 Suspension scheme [47, Fig. 4] for the CubeSat solar array deployment test shown in Fig. 3.10.

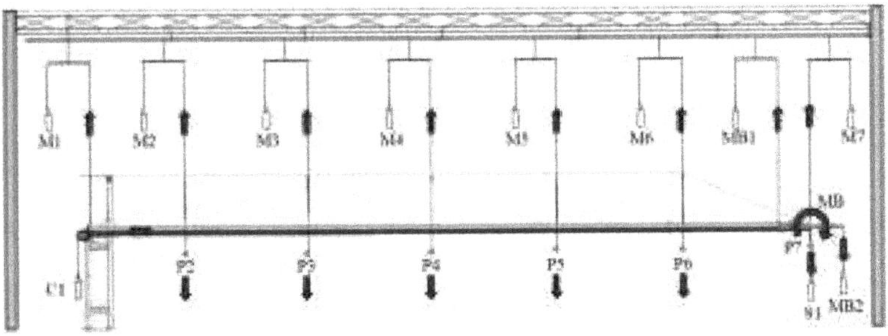

Fig. 3.12 Schematic of solar sail boom deployment test counterweights [26].

just highlighted, fly beams and pulleys are mechanically equivalent. A kinematic difference is that the actual effective span of a fly beam varies with the beam's tilt as the specimen moves; this is not the case with a pulley solution.

A simple example of a pulley-based counterweight offloading has already been shown in, and discussed in connection with, Fig. 3.4. A classic example of fly beam application is schematically presented in Fig. 3.12 for a solar sail boom deployment test [26]. Here, with an added level of complexity touched upon later in this chapter, the fly beams are suspended from carts that move along rails as the specimen deploys—the boom extends, rightward in Fig. 3.12.

Another, less typical, fly beam application for the deployment testing of a Z-folded solar array [5, 16] is shown in Fig. 3.13. An interesting aspect of this example is that the system mass isn't doubled by the counterweights because the specimen itself is used as the counterweight: Each fly beam links one support point to the other, adding no extra inertia beyond that of the beam and the cords. In this example also, the fly beams move along on rails as the specimen deploys.

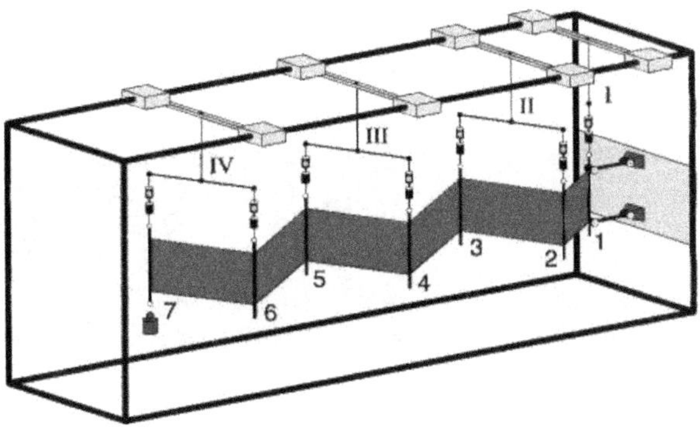

Fig. 3.13 Solar array deployment test setup, from figures in Refs. [5, 16].

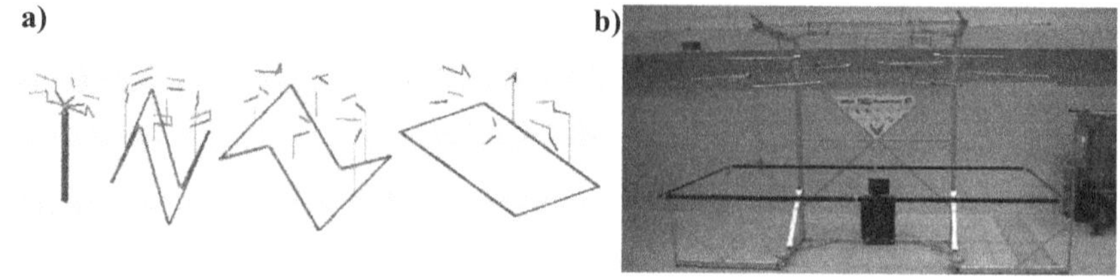

Fig. 3.14 Deployment test for PowerSail unit [8]. (a) Deployment and support scheme. (b) Deployed hardware.

The deployment test of one of the AFRL PowerSail space solar array units, shown in Fig. 3.14, has yet another noteworthy feature: some fly beams hang from others, constituting a partially hierarchical structure. However, in the mechanical sense, this setup isn't substantially different from that shown in Fig. 3.13: both are mechanisms of load distribution, minimizing added inertia.

As the ultimate step of this conceptual progression, a specimen can be suspended with a single hierarchical fly beam system (whiffle tree) from one (and only one) external support. In this specific case, the mechanical characteristics of the suspension system acquire a remarkable feature [4]: if the hierarchy hangs from a *single* point, the mechanical constraints it imposes on the specimen are truly equivalent to weightlessness. The only constraint imposed is the stationary state of the test article's center of gravity—the very constraint associated with a free-floating system (due to the preservation or momentum). As shown in Fig. 3.15, this constraint doesn't depend on the internal structure (topology and geometry) of the support assembly.

However, for the weightlessness-equivalence just described to occur, the criterion of whiffle tree support from a single-point suspension must be strictly enforced, in the sense that a single fly beam hierarchy alone should support the test article. If further support (even if another fly beam hierarchy) is added, the suspension will no longer be mechanically equivalent to the free-floating state and the role of the whiffle tree(s) will be reduced to simple load distribution. An example of such support is shown in Fig. 3.16.

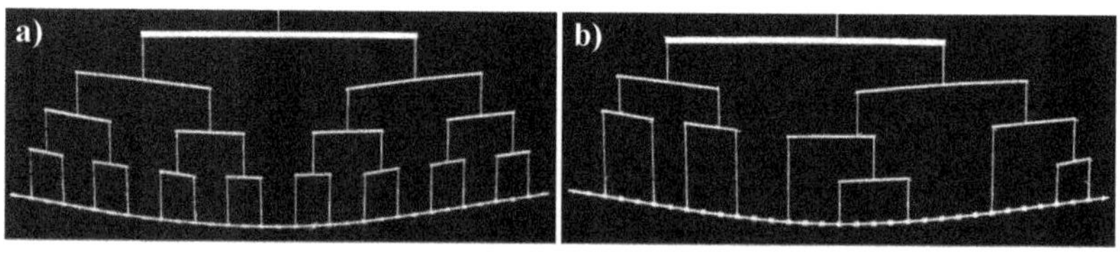

Fig. 3.15 Equivalence with the free-floating state is unconditional for any hierarchical fly beam suspension system hung from only one external point, independent of its internal structure [4, Fig. 11]. (a) Symmetric load tree. (b) Asymmetric load tree.

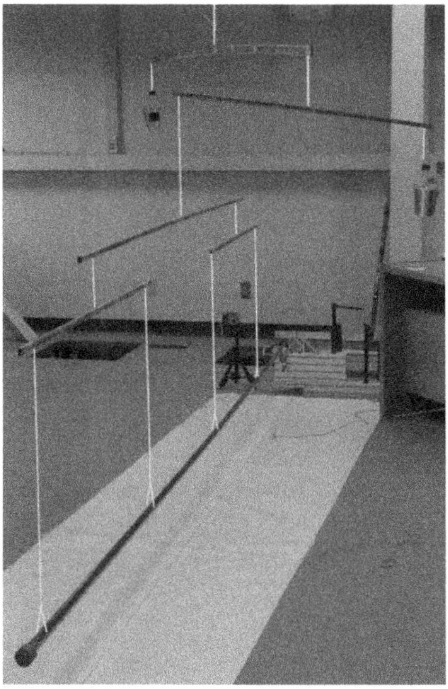

Fig. 3.16 Fly beam hierarchy supporting a boom for buckling test, [48, Fig. 15a] with the boom far end also clamped. Full equivalence to weightlessness (not necessary for the test performed) would be attained if the end clamping were removed.

Beyond the disadvantages already mentioned, two additional limitations of counterweight and weight balancing support are dynamic sensitivity to vibration in the suspension and counterweight system, and the need for space above the specimen. Both drawbacks can be mitigated if compliant support is realized not by mechanically "pure" weight balancing, but with spring- and mechanism-based compliant support, as described next.

Compliant Support via Mechanical Means

To overcome some of the limitations of weight balancing support yet to preserve (or, at least, approximate) their zero stiffness, near-constant offloading forces can be provided with simple mechanisms that reduce the effective elastic constant of an embedded spring.

Such schemes generally employ rotational ("clock") springs which generally feature compact size coupled with a spring constant lower than their linear counterparts. One way to neutralize even the small clock spring constant is to lead the cable that couples the spring to the test article through a pulley or "winch" of non-uniform radius [49, p. 184]; see also Refs. [50–52]. Via the proper pulley profile, the effective support force can be rendered constant (stiffness can be eliminated) for a range of the motion. The concept is illustrated in Fig. 3.17.

Another option to reduce clock spring stiffness is to wrap the spring tape also about a second drum, in the reverse direction [3], as shown in Fig. 3.18. The concurrent reverse bending of the power tape on the two drums results in a very nearly constant resultant over a considerable range of motion, with little parasitic inertia. Drum radii, and the radius of the attached pulley, can be tailored to achieve the desired force.

Ways to reduce the apparent stiffness of a suspending linear spring are also available. For example, the rate effect of a suspending primary spring can be compensated with a secondary one, coupled to the former with a simple mechanism [28, 53].

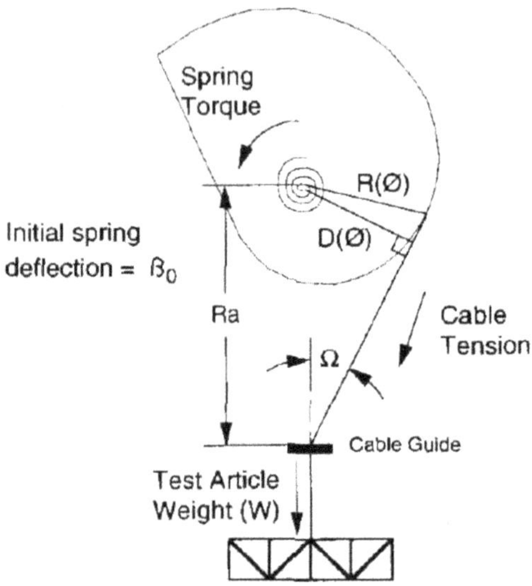

Fig. 3.17 Variable radius cam to neutralize a rotational spring constant [52, Fig. 3].

Pneumatic Support

Pneumatic, or pneumatically assisted, support is a state of the art offloading option, with decades of heritage in high-end, quality, testing. Such support was briefly summarized in the "Pneumatic Systems" section, earlier in this chapter.

Support Movement along Rails

Evolving specimen configurations are often accommodated by moving the external support units, whether the latter hold the specimen from below or above. Commonly, this is accomplished with carts on rails. Offloading examples involving such solutions have already been discussed (see, for example, Figs. 3.12 and 3.13).

A further example for rail-guided support motion and also for a noteworthy use of cables and pulleys [54] is shown in Fig. 3.19: a pulley system integrated with a lightweight trolley on a rail in a manner to permit the counterweight to remain horizontally stationary against the backdrop of horizontal specimen kinematics. This solution alleviates test inertial pollution because only the lightweight cart needs to follow the expanding specimen.

This example is typical [8, 55] in that the cart moves along the rails passively. If friction between rail and trolley is small (this can be achieved, for example by bearings/wheels), the deploying structure can drag the carts along, with the suspending cables nearly vertical. This cord alignment limits second order stiffness effects (which work against the deployment) without the need for extreme suspension heights. However, passive physical coupling of the carts with a test article adds the cart's inertia to the specimen in the lateral direction.

The sliding or rolling of the supports itself can dynamically perturb the test. For sensitive problems, stiction of bearings or between rail and sliding trolley can be a problem, adding excitation to a suspension system that is dynamically hypersensitive to begin with. Stiction was observed, for example, during the testing of AFRL's PowerSail [8, 55] when offloading carts passively slid on rails.

Stiction as well as frictional and geometric nonlinear drag effects can be mitigated or entirely eliminated by actively moving the carts or sliders; this can be achieved in any convenient way—for example, by pulling the carts with

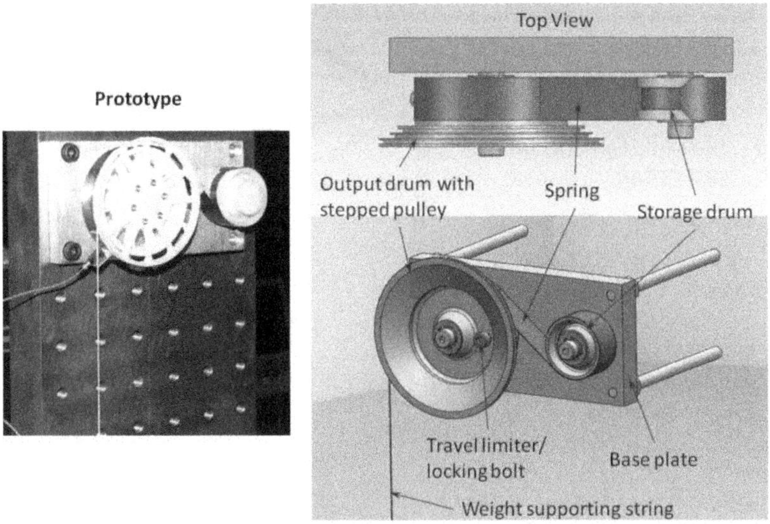

Fig. 3.18 Bi-directionally wound clock spring in long-stroke offloading device [52, Fig. 3].

cords [56]. In some test programs, tests with passive and active tracking have both been performed subsequently for the same specimen [11, 57].

An interesting example of active trolley motion is a motorized cart to autonomously follow the specimen, discussed in Ref. [58]. In this example, the motor is activated by the lead angle of the suspending cable: once the cord tilt reaches 5 deg off the vertical, the motor turns on and moves the cart onward. Such a simple, autonomous, solution also invites control challenges, as is also reported in Ref. [58].

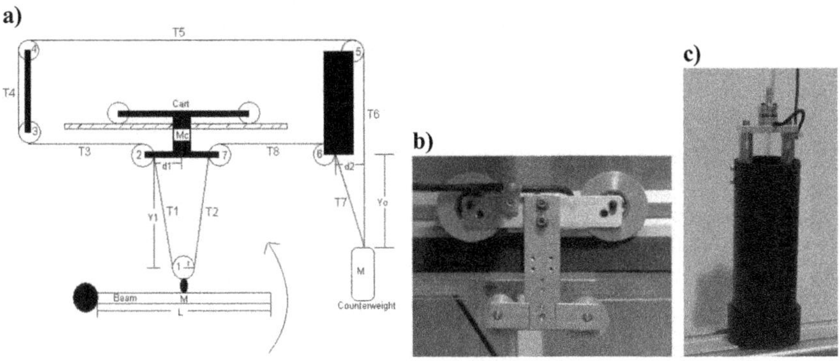

Fig. 3.19 Stationary and rail-moving pulleys to offload moving specimen anchor points with stationary counterweight [54, Figs. 14, 6, 7]. (a) System schematic. (b) Cart on rail. (c) Counterweight.

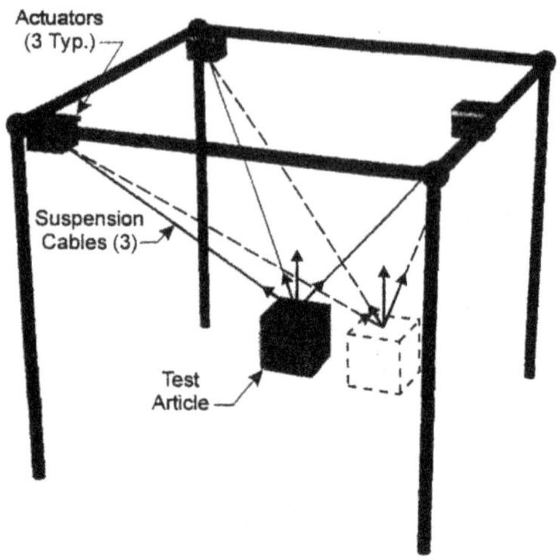

Fig. 3.20 Suspension cable tripod with actively controlled cord release mechanisms [59, Fig. 2].

3.4.3.5 TRUE THREE-DIMENSIONAL SUSPENSION

To accommodate complex spatial motion for test article anchor locations without geometric nonlinear effects associated with off-vertical cable alignment, a support paradigm more general than those reviewed so far is needed. One example for such a solution is the suspension of each anchor point from fixed external locations with a cable tripod such that the release and reeling in of the cords is actively controlled to always provide the desired offloading force. Such a scheme is shown in Fig. 3.20.

For such a setup to work, sufficient unobstructed space above each specimen anchor point is needed. Consequently, the anchor points cannot be too closely spaced unless the motion allows the gross operational space of the cable tripods to overlap. This constraint may severely limit applicability in the case of highly compliant structures which require a number of suspension points.

3.4.3.6 POINT-BY-POINT AUTONOMOUS SUPPORT

Truly three-dimensional motion can also be supported from below, by "holding up" the specimen. To this end, offloading units can be moved "freely" via autonomous units, following the evolution of a specimen configuration. This is the world of innovative "puppeteering" solutions and special engineered compromises, making the best use of available resources within the constraints of budgets and of the physical limits of how various supports can be moved and manipulated.

To illustrate this paradigm, Fig. 3.21 shows a unit of a compound gravity compensation system for variable flexible spacecraft: one of several mobile robots that control support forces and move along with the support points of the test article [60]. The robots use autonomous control to accomplish their task.

3.4.3.7 MODIFYING THE PROBLEM OR THE SPECIMEN

Finally, there are deployment sequences for which none of the above schemes work "out of the box" because there are phases of the deployment during which the tested structure itself blocks access to the locations where support could be applied.

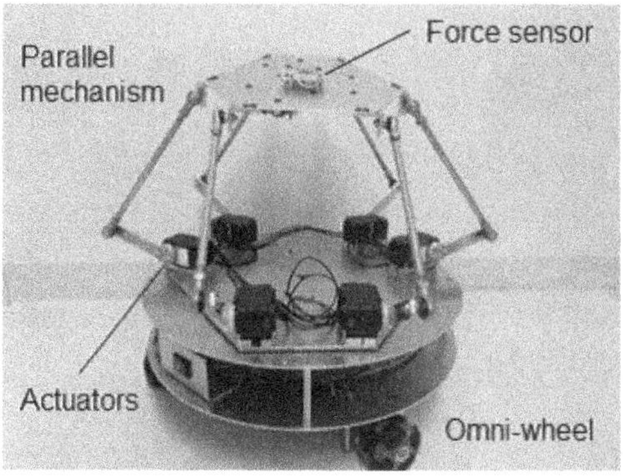

Fig. 3.21 Member of robotic fleet, to autonomously move along deploying hardware while supporting it with the required force [60, Fig. 7].

Two possibilities may be considered in such a case as a final resort. One is to follow the process "piecewise," with different support configurations and, possibly, even different hardware models for each phase. (This approach will work only for quasi-static deployment sequences.)

The other option is to modify the test hardware itself. This usually means that auxiliary structures, typically appendages, are added to permit an acceptable grip. One example is the concept demonstration scaled testing [61, Fig. 1] of the Membrane Optical Imager Real-time Exploitation (MOIRE) $D = 10$ m primary diffractive optics. The structural design of this element is a flat circular array of 18 petals, stowed into a circular "flower bud" shape and deployed as shown in Fig. 3.22.

Because the deployment kinematics didn't permit the suspension of the petals at the centers of gravity (this would have required the suspension cables to pass through the petals when stowed), the petal CGs were moved outboard by adding a counterweight to each petal. To this end, aluminum rods with steel pipes and lead weights at the ends were used, so that cable anchor points on the rod, just outside where the latter protruded from the petal, could be used as shown in Fig. 3.23. More about this test program is discussed in the "MOIRE Primary Optics Deployment Testing" section, later in this chapter.

3.4.3.8 COMBINATION SCHEMES

Finally, the potential for configuration complexity is greatest when different offloading approaches are combined.

At this point, however, it must be noted that "complexity" (the quality guiding the present narrative) is not referred to here with the negative overtones this term often carries in engineering, especially for space systems. It is understood that the best solution strikes the simplest possible balance between project objectives. A less than trivial answer to any challenge is still the simplest means that can be afforded against difficult constraints. Besides, it is not without ambiguity how the term "complex" is used on these very pages. For example, the pneu-mag system, the best solution for many offloading problems, is really a combination scheme coupling pneumatic load bearing with electromagnetic fine tuning, yet it was discussed earlier due to its externally simple modular design.

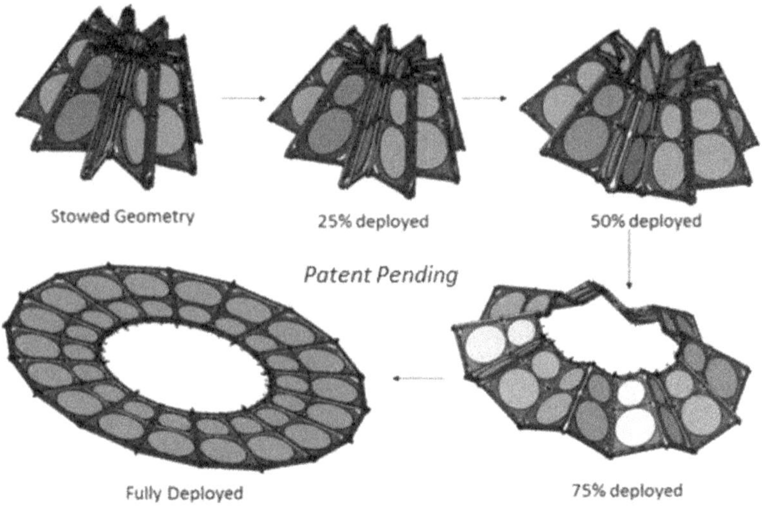

Fig. 3.22 MOIRE primary optics deployment concept [61, Fig. 2].

Accordingly, the combination schemes alluded to in the present subsection reflect a classification of limited scope without value judgment. For an example for such a measure, see the combination of pneu-mag vertical support with flat air bearings for unconstrained horizontal motion described in Ref. [1].

3.5 SOME COMMON APPARATUS ELEMENTS

This section highlights some occasionally under-appreciated aspects of a few simple hardware components that are often used in gravity offloading solutions.

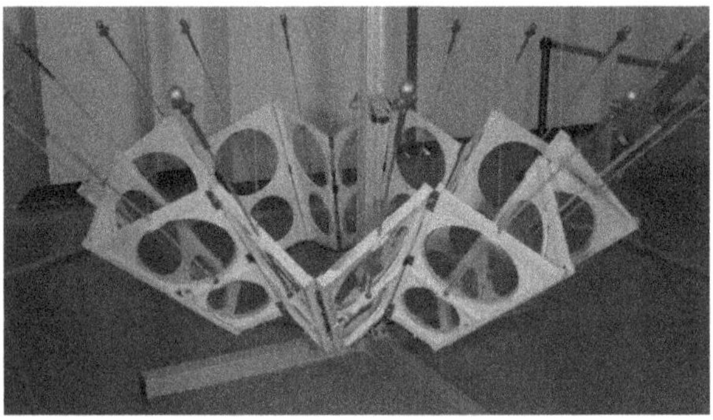

Fig. 3.23 Primary optics scale model during deployment. View with the petal appendages [62] (compare to Fig. 3.27).

In particular, attention is directed toward cables, rollers, fly beams (fly rods), and test frames. Although this list of "lowly" components could be extended (to include sliding/contact elements, tools for quick and simplified measurements and estimates, attachment fixtures and fasteners, etc.), it was decided that only most basic parts with the most general and fundamental utility be addressed.

3.5.1 CABLES AND CORDS

Tension lines in test apparatus come in all types and colors, from simple sewing threads through fishing lines [47] through high-strength cords such as Dacron and Dyneema to high-tech fibers such as Kevlar and Vectran [3]. By virtue of their superior structural efficiency, lines are the lightest, most "ethereal" in most assemblies (unless their stiffness needs to be very high) and, since they rarely have to bear high loads, they are often paid little attention to. However, their structural efficiency is due to a lack of redundancy which, in turn, means a higher risk for failure. Accordingly, when selecting the right cord, one should keep in mind a few practical points.

First and foremost, ensure cord strength is sufficient. Testing this parameter is trivial, but one should not forget that the expected loads are not necessarily limited to those experienced during testing. These loads can also include handling and accidental loads that can easily exceed the former by an order of magnitude. Next, remember that a knot can effectively reduce cord strength to about a third with respect to the smooth state. In addition, possible wear on non-smooth or abrasive surfaces (such as on screw threads or unpolished metal) further reduces strength. Considering these effects, and noting that, by virtue of their efficiency, tension lines are a generally negligible mass overhead on a test article, one should boldly oversize cords. Remember that, from the viewpoint of system optimization, greater benefits are expected from reducing the overheads by components of lesser efficiency!

If cord strength may be an issue, the lines should be tested also after their installation in a setup, in a controlled manner. Via simple means (such as with dummy loads) they should be subjected to the testing forces, to perturbations (such as when the test article is intentionally excited or is accidentally bumped into), handling loads (such as during test specimen removal and installation), and catastrophic events (such as the snapping of one of several tension lines). It is not worth it to risk an expensive, elaborate structure and, possibly, human safety, by overlooking basic steps of precaution.

A key practical concern is how easy and safe it is to attach cords to each other or to other hardware. For example, some stiff fishing lines, Kevlar, and thin wires are hard to tie knots on without damaging them, and clamping a cord under a screw may actually cut the cord. Common sense should be followed and, possibly, auxiliary fixtures for cable management may be needed.

It can be a challenge to manage suspension cables. This can be the case even for structures that, by themselves, are not dominated by cord and cable elements [55, 63].

Beyond strength and practical hands-on issues, stiffness can also be a concern, in two entirely different contexts. One context relevant to cable stiffness is that of the test itself. Cord vibration only in the lateral direction is usually an issue, and this phenomenon doesn't depend on cable stiffness. However, elastic cord elongation can in rare cases also interfere with specimen response. To not affect transient or dynamic response, line stiffness should render stretch under the inertial loads insignificant. The less inertial and stiffness pollution an apparatus imposes on the tested structure, the more likely this may be an issue. (Such is the case, for example, when the specimen weight is offloaded by itself [4].)

The other context in which cable stiffness is relevant is the setup of quasi-static deployment tests where compliant lines are often preferred because they are more "forgiving" and may be easier to work with. Soft cords (such as woven strands like some Dacron leaders) can extend extensively. When preparing an experiment, one should account for their future length change.

3.5.2 ROLLERS AND PULLEYS

Rollers, pulleys, and wheels lead cables and reduce friction for various components' motion on rails or on surfaces. However, the selection of these components for any particular test setup is rarely a trivial step because the common applications for which off-the-shelf, low-cost rolling elements are engineered rarely match the needs of gravity compensation.

Often, the actual load involved differs a lot from what an off-the-shelf rolling part may have been designed for. This is unfortunate because friction in such a component is engineered to be minimal for a certain load range; a quality heavy-duty roller may operate under low loads with much higher apparent friction than a poor-quality lightweight alternative. It may take much trial and error until the right part is selected. The search may take one even to toy stores and thrift stores, until some rolling component of desirable qualities is identified. However, when the search is over, the final selection may still only be a compromise, given various constraints and circumstances.

If no off-the-shelf option works, which is often the case, one may have to fabricate the component from simple elements. Options include ball bearings, shoulder screws, bushings, Teflon or nylon washers, spacers, or bearings, as well as parts that are lathed, perhaps from a 3-D printed initial block. The selection of proper materials is important not only to reduce mass, but to limit friction. Rolling friction is generally lower between hard surfaces; for practical (fabrication) reasons, the rail is often made of hard materials but the roller itself is of a softer substance. Roller size also matters: A larger-diameter wheel is less sensitive to surface irregularities, facilitating steadier, smoother motion.

3.5.3 FLY BEAMS

A rigid body that channels forces between axial members in a support system, the fly beam (or fly rod) is a very common element. In essence, it offers a trivial mechanical means to "split" a force into components at given locations (alternatively, to reduce forces to their resultant). Any particular distribution pattern for a given load can be achieved with the properly designed recursive application of fly beams. In case the beam system is uniformly hierarchical, it is usually called a whiffle tree.

Although fly beams are one of the most frequently used components in gravity offloading technology, one of their critical characteristics is often overlooked: they normally affect a test not only by their (often small) added inertia, but also by a geometric nonlinear effect that is equivalent to adding a torsional spring [4, 7] to each fly beam fulcrum. For the classic, trivial, fly beam construction—which involves a cord from each arm to hold a specimen (or further suspension elements) and a single cord supporting the rod at the fulcrum—this added stiffness is systematic and inevitable. The effect is inescapable because, for the three forces (along the arm cords and the fulcrum cord) to be in stable equilibrium, a free rod will roll to orient itself with the effective fulcrum-cord attachment point to be above the line between the arm endpoints.

This added stiffness effect can be effectively neutralized if the fly beam fulcrum cable is forked and if its two branches are attached to two sides of the rod, bracketing the idealized fulcrum [4, 7] as shown in Fig. 3.24. Although the geometric stiffness effect for each fly beam due to the net fulcrum offset itself will not disappear (because bipod suspension cannot eliminate a fulcrum offset), the error can be rendered random across the set of fly beams, with its statistical mean tending to zero. This won't happen with classic, single-cord fulcrum suspension because the latter solution renders all added stiffnesses uniform in terms of their sign as the fly rods all roll to their lowest energy orientation.

Another critical issue, better known and typically more severe, is the geometric nonlinear rotational stiffness effected on the specimen by suspension attached to the specimen with an offset from the latter's CG. This is a common issue because one's choice of cable anchor locations on a test article is often severely constrained by what part of the specimen exterior is accessible, given the context of the responses and kinematics expected. As a result, it may be impossible to suspend a specimen without support eccentricity which begets parasitic moments in response to specimen rotations, similar to the fly beam effective fulcrum offset discussed above but on a much higher scale. An elegant and simple answer to this problem is a fly beam design with the cord anchor points

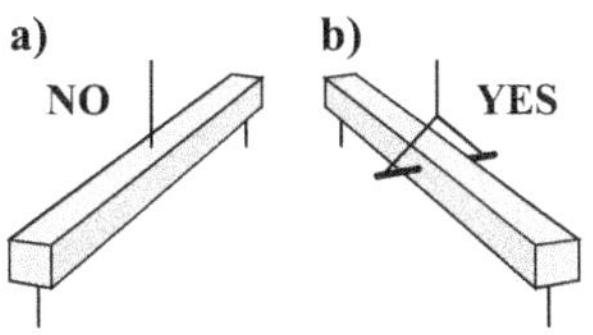

Fig. 3.24 To prevent geometric stiffness effects due to the fly rod's roll into an orientation where the effective fulcrum is above the line between the arm ends, bipod (or equivalent) fly beam suspension is necessary [7, Fig. 13].

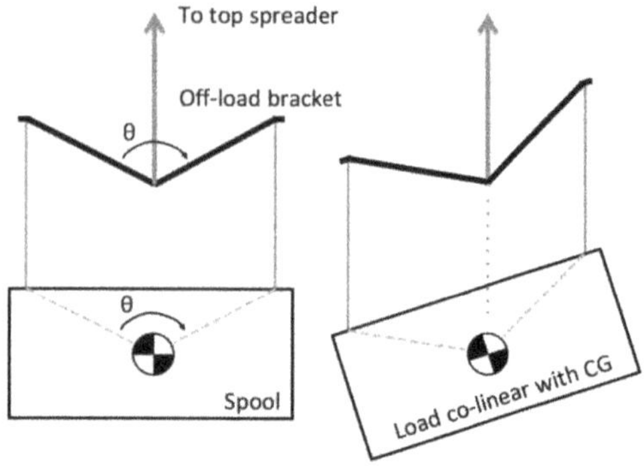

Fig. 3.25 To eliminate geometric stiffness effects due to specimen anchor point placement, the fly beam geometry should reflect that of the test article CG and anchor point locations [64, Fig. 4b].

geometrically reflecting the arrangement of the same cords' anchor point pattern on the specimen and the latter's CG [64]. This principle is illustrated in Fig. 3.25.

3.5.4 EXTERNAL SUPPORT—SUPPORT FRAME

Test frames and rigs have to hold not only the specimen, via a gravity compensation mechanism, but also data-acquisition devices. Such external support often has to be orders of magnitude heavier and stiffer than the structures to be tested. This is especially true when the support rig provides a near-idealized mechanical inertial reference, and when the test measurements have to be precise. Occasionally, support frames as massive as those in civil engineering test laboratories are used.

In some circumstances, the test frame itself has to be placed in a thermally damped and isolated enclosure, the entire system possibly erected within the laboratory.

However, for many initial deployment verification tests (and for some deployed performance tests), a near ideal mechanical reference isn't necessary. This is because, due to specimen compliance and low weight, the precise quantitative reproduction of in-space behavior remains a distant objective regardless of the test frame. In such cases, no more than an *ad-hoc* rig may be needed. For example, one may erect a frame from lumber that is readily available, cheap, and easy to work with. In addition, various commercial modular products can be "borrowed" to serve as a test frame. Modular mechanical shelf systems and some plumbing supplies (such as pipes with straight, elbow, and T-joint elements) can also be used as modular components from which frames can be quickly erected and later torn down, ready for reuse.

The suspension of gravity compensation devices from such *ad-hoc* frames is generally not more difficult than attachment to more solid support, including walls and ceilings. All innovative solutions are welcome, as long as the test itself is not jeopardized.

Finally, note that an *ad-hoc* test setup doesn't necessarily render the acquisition of quantitative (rather than qualitative) information futile. For a sound quantitative evaluation of the results, however, the mechanical behavior of the entire test setup (specimen, gravity compensation apparatus, support frame, and electronic components) needs to be faithfully simulated with numerical-analytical tools.

3.6 DEVELOPMENT STORIES

Few things are more educational to both the uninitiated and the expert than case studies. Although much can be learned about gravity compensation problems and solutions from bits of information collected from miscellaneous publications, a direct narrative of why and how offloading decisions were made and why certain compromises were accepted can be much more illuminating. Further, such a narrative can also shed light on logistical details that can be as important as the principles and solutions themselves.

With a keen eye on the literature, one can identify projects with sufficiently complete publication trails in which the process of gravity compensation methodology development is also described. Such publications, together, often describe complete or nearly complete stories of our current interest.

One such example is the validation testing of the 5.5-m diameter UltraFlex-175 deployable accordion fanfold flexible blanket solar array [65–68] with tests performed both in ambient pressure and in vacuum and attention paid both to deployment itself and to deployed performance. The publication track also reveals how the offloading approach changed during years of development. Several intriguing technical details are discussed, including discrepancies between tests and analyses, the diagnostic evaluation of observed mismatch, efforts to eliminate the mismatch, and the methodology to finetune the surrogate numerical analysis procedure.

Another example of an insightful offloader development story is associated with the development and testing of a Japanese mesh antenna a decade ago. Although the applied structural concept was conceived to minimize the effect of gravity on the surface itself by virtue of a modular surface structure [57, 69], problems with the offloading system for the skeletal structure as a whole were still encountered and subsequently remedied [9, 11].

The "MOIRE Primary Optics Deployment Testing" section describes the development story of the offloader system for the deployment concept verification model of Ball's MOIRE Primary Optics—courtesy of Phil Aumiller, Principal Mechanical Design Engineer at Ball Aerospace.

3.6.1 MOIRE PRIMARY OPTICS DEPLOYMENT TESTING

The objective of the MOIRE (Membrane Optical Imager Real-time Exploitation) program, spearheaded by Ball Aerospace, is a 50-m system with $D = 10$ m primary diffractive optics (to be later traced to twice this size), as illustrated in

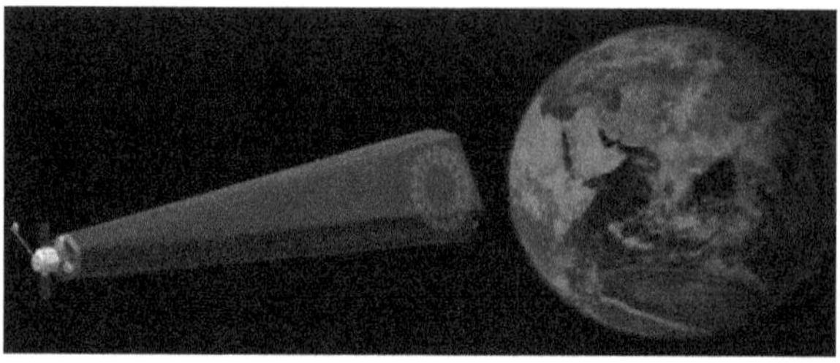

Fig. 3.26 Artist's concept of diffractive optical system [61, Fig. 1].

Fig. 3.26. The primary optic structural design consists of a flat circular array of 18 petals, stowed into a circular "flower bud" shape and deployed as was shown in Fig. 3.22.

Early in the project, it was necessary to demonstrate that deployment was possible without active control. Accordingly, the decision to build a 1/6 scale model was made. With the deployment force supplied by spring hinges between the petals, the assembly had to be supported, allowing the spring hinges to push the petals to their deployed flat configuration without gravity aiding or hindering the deployment.

The first idea, to suspend each petal from its center of mass, proved impractical because it would have required the suspension cables to pass through the petals when in the stowed position. The solution was to move the petal CGs outboard by adding a counterweight to each petal. To this end, aluminum rods with steel pipes and lead weights at the ends were used. Each rod was attached to the petal with ball bearings to allow free rotation. The counterweight was adjusted to

Fig. 3.27 MOIRE primary optics scale model deployment test offloading setup. Compare to Fig. 3.23. (Courtesy of Ball Aerospace, Phil Aumiller.)

Fig. 3.28 Triangular fly beams and counterweights, MOIRE primary optics scale model deployment testing. (Courtesy of Ball Aerospace, Phil Aumiller.)

balance the petal assembly about the cable anchor point on the rod, just outside where the latter protruded from the petal, as was shown in Fig. 3.23.

To maintain vertical orientation for the cables during deployment, they were suspended from linear bearings on precision-ground stainless steel rods arranged in an even radial pattern. However, the motion during deployment of the petal/counterweight CGs wasn't precisely radial. Because the inter-petal hinge fold directions alternated (the hinges folded in and out, repetitively), vertical cable alignment in stowage implied uneven cable spacing, with gaps of about 1 and 3 in. On the other hand, cable spacing had to be even in the deployed state. This change was accommodated by inserting a "yoke" (a small fly beam) into each suspending cable, with the fulcrum, at the top, suspended from the linear bearing, one arm suspending the petal rod, and the other a weight equal to the combined weight of petal and its counterweight on the rod, as shown in Fig. 3.27.

In fact, the yokes were triangular, with the fulcrum markedly above the arm anchor points, as shown in Fig. 3.28. This shape made the yokes self-balancing at the cost of a local geometric nonlinear torsional stiffness perturbation, eliminating the need for precision for the counterweights. As the array of petals deployed, the yoke pivoted, maintaining vertical alignment for all cables.

This offloader design, built in about two months, was a low-tech project for a high-tech problem. The rods were supported by a modular structure of extruded aluminum beams bolted together and industrial railing fittings. The counterweights were lead fishing weights. The spring hinges, bearings, cables, and fittings were off-the-shelf, low-cost components. The petal frames were made from a structural foam plastic, cut with a CNC (Computer Numerical Control) router by the plastic supplier. The yokes were custom made from sheet metal. The pieces to hold the bearings were 3D printed. The deployment mechanism used fishing line and a remote-operated commercial, low-cost actuator.

The end result was an affordable model that successfully demonstrated what was intended, readily scalable to full size if needed. To gauge the robustness of the design, problems such as uneven hinge spring forces and added friction/drag were also simulated.

ACKNOWLEDGMENTS

The writer of this chapter gratefully acknowledges the help of many colleagues who helped his work by sending him references, answering questions, or assisting with parts of the composition. Special help is due to Steve White of Deployable Space Systems, Inc., who proposed and initially outlined much of the core material for the "Golden Rules" section; to Phil Aumiller of Ball Aerospace for writing a draft for the description of Ball's MOIRE primary optics deployment verification testing and for sharing photographs; to Andrew Viquerat and Gabriel Secheli of the University of Surrey for sharing photos and practical information about balloon-offloaded testing programs; to Thomas Murphey of Opterus R&D for sharing unpublished references and for insightful conversations; and to Joe Footdale of LoadPath, Inc., for insightful conversations. The help of Richard Pappa of NASA Langley R. C., who provided references and photographs from past programs, is also gratefully acknowledged.

REFERENCES

[1] Kienholz, D. A., "Simulation Of The Zero-Gravity Environment For Dynamic Testing Of Structures," The 19th Space Simulation Conference Cost Effective Testing for the 21st Century, IES/AIAA/ASTM/NASA/CSA, NASA, Goddard Space Flight Center, National Technical Information Service, Baltimore, MD, Oct. 28–31, 1996, pp. 219–230, SuDoc NAS 1.55:3341, NASA-CP-3341, Rept-97A00653. Jan. 1997.

[2] Han, O., Kienholz, D., Janzen, P., and Kidney, S., "Gravity-Offloading System for Large-Displacement Ground Testing of Spacecraft Mechanisms," Proc. 40th Aerospace Mechanisms Symposium, May 12–14, 2010, at Hilton Cocoa Beach Oceanfront Hotel, Cocoa Beach, FL, Hosted by the Kennedy Space Center and Lockheed Martin Space Systems Company, Mechanisms Education Association, May 12–14, 2010.

[3] Ardelean, E., Jeon, S., and Cooper, B., "Dynamic Behavior of a Low Inertia Gravity Off-load Passive Device," Ref. [70], AIAA 2012-1954, DOI 10.2514/6.2012-1954.

[4] Greschik, G., and Belvin, W. K., "A High-Fidelity Gravity Offloading System for Free-Free Vibration Testing," *Journal of Spacecraft and Rockets*, Vol. 44, No. 1, Jan.–Feb. 2007, pp. 132–142.

[5] Fischer, A., and Pellegrino, S., "Interaction Between Gravity Compensation Suspension System and Deployable Structure," *Journal of Spacecraft and Rockets*, Vol. 37, No. 1, Jan.–Feb. 2000, pp. 93–99.

[6] Schultheiss, D., "Gravity Compensation of Deployable Solar Arrays for Small Spacecraft," *Diploma Thesis for the degree of Diplom-Ingenieur, Universitat Stuttgart Institut fur Statik und Dynamik*. By the University of Stuttgart, Germany, and the University of Cambridge, UK, Department of Engineering, Nov. 2003.

[7] Greschik, G., "Marionette Suspension in the Testing of Long Light Structures," Ref. [71], AIAA 2007-1821, DOI 10.2514/6.2007-1821.

[8] Adler, A., Hague, N., Spanjers, G., Engberg, B., Goodding, J., David, D. M., and Mikulas, M. M., "PowerSail: The Challenges of Large, Planar, Surface Structures For Space Applications," Ref. [72], AIAA 2003-1828.

[9] Meguro, A., Shintate, K., Usui, M., and Tsujihata, A., "In-Orbit Deployment Characteristics of Large Deployable Antenna Reflector On-board Engineering Test Satellite VIII," 56th International Astronautical Congress, International Academy of Astronautics, IAA, Hyderabad, India, Sept. 24–28, 2007.

[10] McEachen, M. E., and Trautt, T. A., "Confirmation of New Analytics for Ultra-Light Lattice Column Strength Using a 40-m Flight Article," 50th AIAA/ASME/ASCE/AHS/ASC Structures, Structural Dynamics, and Materials Conference 17th AIAA/ASME/AHS Adaptive Structures Conference 11th AIAA Non-Deterministic Approaches Conference 10th AIAA Gossamer Spacecraft Forum 5th AIAA Multidisciplinary Design Optimization Specialist Conference AIAA Dynamics Specialists Conference, AIAA/ASME/ASCE/AHS/ASC, AIAA, Palm Springs Convention Center/Wyndham Palm Springs, Palm Springs, CA, May 4–7, 2009, AIAA 2009-2169, DOI 10.2514/6.2009-2169.

[11] Meguro, A., Ishikawa, H., and Tsujihata, A., "Study on Ground Verification for Large Deployable Modular Structures," *Journal of Spacecraft and Rockets*, Vol. 43, No. 4, July–Aug. 2006, pp. 780–787, DOI 10.2514/1.16037.

[12] Greschik, G., and Mikulas, M. M., "Design Study of a Square Solar Sail Architecture," *Journal of Spacecraft and Rockets*, Vol. 39, No. 5, Sept.–Oct. 2002, pp. 653–661, journal version of paper AIAA-00-2001-1259, presented at the *2001 SDM Conference.*

[13] Greschik, G., Derbes, B., Veal, G., and Rogan, J., "The Cord Mat Sail—Concept, Mechanics, And Design Example," The 46th AIAA/-ASME/ASCE/AHS/ASC Structures, Structural Dynamics, and Materials Conference and AIAA/ASME Adaptive Structures Forum, Technical Papers, AIAA/ASME/ASCE/AHS/ASC, AIAA, Austin, TX, April 18–21, 2005, AIAA-2005-2049.

[14] Sleight, D., Mann, T., Lichodziejewski, D., and Derbes, B., "Structural Analysis and Test Comparison of a 20-Meter Inflation-Deployed Solar Sail," Ref. [73], AIAA 2006-1706, DOI 10.2514/6.2006-1706.

[15] Menon, C., Busolo, S., Cocuzza, S., Aboudan, A., Bulgarelli, A., Bettanini, C., and Angrilli, F., "Issues and New Solutions for Testing Free-Flying Robots," 55th International Astronautical Congress of the International Astronautical Federation, the International Academy of Astronautics, and the International Institute of Space Law, International Astronautical Federation, 3–5, Rue Mario-Nikis, 75015 Paris, France, IAF, Vancouver, Canada, Oct. 4–8 2004, IAC-04-J.5.08.

[16] Fischer, A., *Gravity Compensation of Deployable Space Structures*, Ph.D. thesis, Pembroke College, the University of Cambridge, Cambridge, UK, Nov. 2000.

[17] Adams, D., *Life, the Universe and Everything*, Pan Books, the UK, 1982, *The Hitchhiker's Guide to the Galaxy*, Vol. 5.

[18] NASA's Glenn Research Center in Cleveland, Ohio, Fact sheet, PS00566141011, 2011, Available at http://facilities.grc.nasa.gov/documents/TOPS/TopZERO.pdf, verified Aug. 22, 2016.

[19] Herr, R. W., and Horner, G. C., "Deployment Tests of a 36-Element Tetrahedral Truss Module," Large Space Systems Technology – 1980. Second Annual Technical Review held November 18–20, 1980, at NASA Langley Research Center, edited by F. Kopriver, III, Vol. 2, NASA Langley Research Center, NASA Scientific and Technical Information Office, Hampton, Virginia, Feb. 1, 1981, pp. 59–69, NASA CP-2168.

[20] Straubel, M., Block, J., Sinapius, M., and Ḧuhne, C., "Deployable Composite Booms for Various Gossamer Space Structures," Ref. [74], AIAA 2011-2023, DOI 10.2514/6.2011-2023.

[21] Chung, D. T., and Young, L. E., "Zero Gravity Testing of Flexible Solar Arrays," Proc. 15th Aerospace Mechanisms Symposium, May 14–15, 1981, at NASA George

C. Marshall Space Flight Center, Alabama, NASA, Wash., D.C., California Inst. of Tech., Pasadena, CA, and Lockheed Missiles & Space Co., Sunnyvale, CA, May 1981.

[22] Kienholz, D. A., "CSA Pneumatic/Magnetic Suspension System Application Note No. 2: Comparison to Buoyancy Devices (Helium Balloons)," *Prepared for Sandia National Laboratories*, CSA Engineering Inc., Palo Alto, CA, May 1993, available from the Moog-CSA website.

[23] Brennen, C. E., "A Review Of Added Mass And Fluid Inertial Forces," *Tech. Rep. CR 82.010*, Naval Civil Engineering Laboratory, Port Hueneme, CA, Jan. 1982, sponsored by Chief Of Naval Material.

[24] Fernandez, J. M., Visagie, L., Schenk, M., Stohlman, O. R., Aglietti, G. S., Lappas, V. J., and Erb, S., "Design and Development of a Gossamer Sail System for Deorbiting in Low Earth Orbit," *Acta Astronautica*, Vol. 103, No. 1, Oct. – Nov. 2014, pp. 204 – 225, DOI 10.1016/j.actaastro.2014.06.018.

[25] Stokes, J., Engler, E., and Agan, W., "Neutral Buoyancy Test Results of a Deployable Space Beam," 2nd Conference on Large Space Platforms: Toward Permanent Manned Occupancy in Space, AIAA, San Diego, CA, Feb. 2 – 4, 1981, pp. 1 – 4.

[26] Lichodziejewski, D., Derbes, W., Reinert, R., Belvin, K., Slade, K., and Mann, T., "Development and Ground Testing of a Compactly Stowed Scalable Inflatably Deployed Solar Sail," The 45th AIAA/ASME/ASCE/AHS/ASC Structures, Structural Dynamics, and Materials Conference and AIAA/ASME Adaptive Structures Forum, Technical Papers, AIAA/ASME/ASCE/AHS/ASC, AIAA, Palm Springs, CA, April 19 – 22, 2004, AIAA 2004-1507, DOI 10.2514/6.2004-1507.

[27] Palisoc, A. L., Redell, F. H., and Andersen, G., "Deployment and Structural Support of Space Membrane Optics System Using Rigidizable Conical Booms," Earth and Space 2004—Proceedings of 9th Biennial International Conference of the Aerospace Division on Engineering, Construction, and Operations in Challenging Environments, edited by R. B. Malla, and A. K. Maji, ASCE, American Society of Civil Engineers, Reston, VA, League City/Houston, Texas, March 7 – 10, 2004.

[28] Kienholz, D., Crawley, E., and Harvey, T. J., "Very Low Frequency Suspension Systems for Dynamic Testing," The 30th AIAA/ASME/-ASCE/AHS/ASC Structures, Structural Dynamics, and Materials Conference and AIAA/ASME Adaptive Structures Forum, AIAA/ASME/ASCE/-AHS/ASC, AIAA, Mobile, AL, April 3 – 5, 1989, AIAA 1989-1194, DOI 10.2514/6.1989-1194.

[29] Kienholz, D. A., "Defying Gravity with Active Test Article Suspension Systems," *SV. Sound and Vibration*, April 1994, pp. 14 – 22.

[30] Moog/CSA Engineering, "Zero Gravity Simulation," Product brochure, electronically available, Nov. 2010 [retrieved on Feb. 24, 2017, from the Moog/CSA website www.csaengineering.com under *Products and Services/Zero Gravity Simulation*].

[31] Greschik, G., Mikulas, M. M., and Freeland, R. E., "Scale model testing of nonlinear phenomena with emphasis on thin film deployable structures," IUTAM-IASS Symposium on Deployable Structures: Theory and Applications, edited by S. Pellegrino, and S. D. Guest, Vol. 80 of *Solid Mechanics and its Applications*, International Union of Theoretical and Applied Mechanics, Kluwer Academic Publishers, the Netherlands, 2000, pp. 127 – 136, Proceedings of the IUTAM Symposium held in Cambridge, UK, Sept. 6 – 9, 1998.

[32] Greschik, G., Mikulas, M. M., and Freeland, R. E., "The nodal concept of deployment and the scale model testing of its application to a membrane antenna," The 40th

AIAA/ASME/ASCE/AHS/ASC Structures, Structural Dynamics, and Materials Conference and AIAA/ASME Adaptive Structures Forum, Technical Papers, Vol. 4, AIAA/ASME/ASCE/AHS/ASC, AIAA, St. Louis, MO, April 12–15, 1999, pp. 2546–2554, AIAA Paper 1999-1523.

[33] Barenblatt, G. I., *Scaling, Self-similarity, and Intermediate Asymptotics*, No. 14 in Cambridge Texts in Applied Mathematics, Cambridge University Press, 40 West 20th Street, New York, NY 10011-4221, Dec. 1996, Series editor: D. G. Crighton.

[34] Salama, M., White, C., and Leland, R., "Ground Demonstration Of A Spinning Solar Sail Deployment Concept," The 42nd AIAA/ASME/- ASCE/AHS/ASC Structures, Structural Dynamics, and Materials Conference and AIAA/ASME Adaptive Structures Forum, Technical Papers, AIAA/ASME/ASCE/AHS/ASC, AIAA, Seattle, WA, April 16–19, 2001, AIAA 01-1266.

[35] Furuya, H., and Inoue, Y., "Dynamic Properties of Rotationally Skew Fold Membrane for Spinning Solar Sail," Ref. [73], AIAA 2006-1980, DOI 10.2514/6.2006-1980.

[36] "Private communication with Dr. Andrew Viquerat and Dr. Gabriel Secheli," May 2015.

[37] Anon., "Fabrication Feasibility Study of a 30 Watt/Pound Roll-Up Solar Array, Final Report," *Report No. 652-00101-FR NASA JSC-11443, Fairchild Hiller Space and Electronics System Division*, Germantown, MD, Aug. 15 1968, Prepared for the Jet Propulsion Laboratory of the California Institute of Technology under Contract No. 951969.

[38] Adetona, O., Horta, L., Taleghani, B., Blandin, J., and Woods, K. J., "Vibration Studies of an Inflatable/Rigidizable Hexapod Structure with a Tensioned Membrane," Ref. [72], AIAA-2002-1737, http://dx.doi.org/10.2514/6.2003-1737.

[39] Adetona, O., Keel, L., Horta, L., Cadogan, D., and Sapna, G., "Description of New Inflatable/Rigidizable Hexapod Structure Testbed for Shape and Vibration Control," Ref. [75], AIAA-2002-1451, http://dx.doi.org/10.2514/6.2002-1451.

[40] Schwartz, L. J., Peck, M. A., and Hall, C. D., "Historical Review of Air-Bearing Spacecraft Simulators," *Journal of Guidance, Control, and Dynamics*, Vol. 26, No. 4, July–Aug. 2003, DOI 10.2514/2.5085.

[41] Welch, A., Smith, S., and Main, J., "Experimental Results Regarding Two-Dimensional Deployment of Inflatable Beams," Ref. [72], AIAA 2003-1976, DOI 10.2514/6.2003-1976.

[42] McEachen, M. E., "Validation of Slender Lattice Trusses by Modeling and Testing," *Journal of Spacecraft and Rockets*, Vol. 49, No. 6, Nov.–Dec., 2012, pp. 970–977, DOI 10.2514/1.51301.

[43] Pickett, W. L., Pratt, W. D., Larson, M. L., Redlin, D. M., and Hinkle, J. D., "Testing of Centrifugally Deployed Membrane Dynamics in an Ambient Ground Environment," Ref. [75], AIAA 2002-1241.

[44] Matunaga, S., "New Gravity Compensation Method by Dither for Low-g Simulation," *Journal of Spacecraft and Rockets*, Vol. 32, No. 2, March–April, 1995, pp. 364–369.

[45] Banik, J. A., Murphey, T. W., and Dumm, H.-P., "Synchronous Deployed Solar Sail Concept Demonstration," 49th AIAA/ASME/ASCE/-AHS/ASC Structures, Structural Dynamics, and Materials Conference 16th AIAA/ASME/AHS Adaptive Structures Conference 10th AIAA Non-Deterministic Approaches Conference 9th

AIAA Gossamer Spacecraft Forum 4th AIAA Multidisciplinary Design Optimization Specialist Conference AIAA Dynamics Specialists Conference, AIAA/ASME/ASCE/AHS/ASC, AIAA, Renaissance Schaumburg Hotel & Convention Center, Schaumburg, IL 60173, April 7–10, 2008, AIAA 2008-2213, DOI 10.2514/6.2008-2213.

[46] Banik, J., and Ardelean, E., "Verification of a Retractable Solar Sail in a Thermal-Vacuum Environment," 51st AIAA/ASME/ASCE/AHS/ASC Structures, Structural Dynamics, and Materials Conference 18th AIAA/ASME/AHS Adaptive Structures Conference 12th AIAA Non-Deterministic Approaches Conference 11th AIAA Gossamer Spacecraft Forum 6th AIAA Multidisciplinary Design Optimization Specialist Conference AIAA Dynamics Specialists Conference, AIAA/ASME/ASCE/AHS/ASC, AIAA, Rosen Centre Hotel, Orlando, FL, April 12–15, 2010, AIAA 2010-2585, DOI 10.2514/6.2010-2585.

[47] Olson, G. M., Murphey, T. W., and Thomas, G., "Free Deployment Dynamics of a Z-Folded Solar Array," Ref. [74], AIAA 2011-1739, DOI 10.2514/6.2011-1730.

[48] Fernandez, J., "Advanced Deployable Shell-Based Composite Booms For Small Satellite Structural Applications Including Solar Sails," Proceedings of the Fourth International Symposium on Solar Sailing (ISSS 2017), held January 17–20, 2017, Kyoto Research Park, Kyoto, Japan., Japan Space Forum, Kyoto, Japan, Jan. 17–20, 2017, Published electronically on the conference web site, link verified on March 1, 2017.

[49] Chironis, N. P., *Mechanisms & Mechanical Devices Sourcebook*, McGraw-Hill, Inc., NY, 1991.

[50] Yang, L. F., and Chew, M.-S., "Ground-Based Testing Of the Dynamics of Flexible Space Structures Using Band Mechanisms," *Tech. Rep. NASA-CR-188154*, NASA, May 1, 1991.

[51] Chew, M.-S., Juang, J.-N., and Yang, L.-F., "Suspension Device for Low-Frequency Structures," *United States Patent*, May 4, 1993, Patent Number US 5,207,110.

[52] Straube, T., and Peterson, L. D., "A large motion suspension system for simulation of orbital deployment," The 35th AIAA/ASME/ASCE/-AHS/ASC Structures, Structural Dynamics, and Materials Conference and AIAA/ASME Adaptive Structures Forum, Technical Papers, Vol. 4, AIAA/ASME/ASCE/AHS/ASC, AIAA, Hilton Head, SC, April 18–20, 1994, pp. 1959–1969, AIAA 1994-1568.

[53] Harvey, T. J., "Adjustment of Zero Spring Rate Suspensions," *United States Patent*, June 18, 1991, Patent Number US 5,024,111.

[54] Tanimoto, R., Moore, A., MacDonald, D., Thomas, S., Murray, A., Polanco, O., and Agnes, G., "Model and Test Validation of Gravity Offload System," Ref. [71], AIAA 2007-1790, DOI 10.2514/6.2007-1790.

[55] Murphey, T.W., "Testing of Large, Ultra-Lightweight Spacecraft," Ref. [70], Presented on Thursday, Jan. 14, at Panel Session 319-SCS-4, of the same title.

[56] Hillebrandt, M., Meyer, S., Zander, M., and Hhne, C., "Deployment Testing of the De-Orbit Sail Flight Hardware," Ref. [76], AIAA 2015-0434, DOI 10.2514/6.2015-0434.

[57] Meguro, A., "A Module Concept for a Cable-Mesh Deployable Antenna," Proc. 27th Aerospace Mechanisms Symposium, May 12–14, 1993, at NASA ARC, Moffett Field, CA, sponsored by NASA, California Inst. of Tech., and Lockheed Martin Space Systems Company, Mechanisms Education Association, May 12–14, 1993, pp. 51–66.

[58] Bihl, T. J., Pham, K. D., and Murphey, T.W., "Modeling and Control of Active Gravity Off-Loading for Deployable Space Structures," Proc. SPIE 6555, Sensors and Systems for Space Applications, edited by R. T. Howard, and R. D. Richards, Society of Photo-Optical Instrumentation Engineers, SPIE, Orlando, FL, April 9, May 29, 2007, DOI 10.1117/12.717410.

[59] Hasselman, T., and Anderson, M., "Development of a Large Amplitude 3D Microgravity Suspension System," The 38th AIAA/ASME/ASCE/-AHS/ASC Structures, Structural Dynamics, and Materials Conference and AIAA/ASME Adaptive Structures Forum, Technical Papers, AIAA/-ASME/ASCE/AHS/ASC, AIAA, Kissimmee, FL, April 7–10, 1997, AIAA 1997 1566, DOI 10.2514/6.1997-1566.

[60] Tsujita, K., Shigematsu, H., and Kishimoto, N., "Development of a Gravity Compensation System for the Prototype Test of Spacecraft by Using Mobile TypeMulti Robots," 12th International Conference on Motion and Vibration Control, Japan Society of Mechanical Engineers, JSME, Sapporo, Hokkaido, Japan, Aug. 3–7, 2014.

[61] Waller, D., Campbell, L., Domber, J. L., Putnam, D., and Thompson, R. T., "MOIRE Primary Diffractive Optical Element Structure Deployment Testing," Ref. [76], AIAA 2015-1836, DOI 10.2514/6.2015-1836.

[62] Ball Aerospace, "Large Diffractive Telescope (MOIRE)," Video, Feb. 9, 2015, https://www.youtube.com/watch?v=QAv06I10Bvw, verified Aug. 15, 2016.

[63] Lane, S. L., Murphey, W. T., and Zatman, M., "Overview of the Innovative Space-Based Radar Antenna Technology Program," *Journal of Spacecraft and Rockets*, Vol. 48, No. 1, Jan.–Feb., 2011, pp. 135–144.

[64] Fulton, J., Jeon, S., and Murphey, T. W., "Flight Qualification Testing of a Meter-class CubeSat Deployable Boom," SciTech 2017: The 4th AIAA Science and Technology Forum and Exposition, AIAA, AIAA, Gaylord Texan Resort & Convention Center, Grapevine, Texas, Jan. 9–13, 2017, AIAA 2017-0621, DOI 10.2514/6.2017-0621.

[65] Spence, B., White, A., Wilder, N., Gregory, T., Douglas, M., Takeda, R., Mardesich, N., Peterson, T., Hillard, B., Sharps, P., and Fatemi, D., "Next Generation UltraFlex Solar Array for NASA's New Millennium Program Space Technology 8," Proceedings of the 2005 IEEE Aerospace Conference, Institute of Electrical and Electronics Engineers (IEEE), IEEE, Big Sky, MT, March 5–12, 2005, DOI 10.1109/AERO.2005.1559374.

[66] White, S., Spence, B., Trautt, T., and Cronin, P., "UltraFlex-175 on Space Technology 8 (SR8) — Validating the Next-Generation in Lightweight Solar Arrays," NASA Science Technology Conference 2007 (NSTC 2007), NASA Glenn Research Center; Cleveland, OH, National Aeronautics and Space Administration Earth Science Technology Office, University of Maryland University College, East Adelphi, MD, USA, June 19–21, 2007.

[67] Eacret, D. L., and White, S., "ST8 Validation Experiment: Ultraflex-175 Solar Array Technology Advance: Deployment Kinematics and Deployed Dynamics Ground Testing and Model Validation," Ref. [77], AIAA 2010-1497, DOI 10.2514/6.2010-1497.

[68] Trautt, T. A., and White, S. F., "ST8 UltraFlex-175 Solar Array — Deployed Dynamics Analytical Modeling and Comparison to Validation Criteria," Ref. [77], AIAA 2010-1498, DOI 10.2514/6.2010-1498.

[69] Harada, S., Meguro, A., and Watanabe, M., "A High Precision Surface Shape Design for Large Deployable Mesh Antenna," Ref. [72], AIAA 2003-1497.

[70] AIAA/ASME/ASCE/AHS/ASC, 53rd AIAA/ASME/ASCE/AHS/ASC Structures, Structural Dynamics, and Materials Conference 20th AIAA/-ASME/AHS Adaptive Structures Conference 14th AIAA Non-Deterministic Approaches Conference 13th AIAA Gossamer Spacecraft Forum 8th AIAA Multidisciplinary Design Optimization and Specialist Conference, Sheraton Waikiki, Honolulu, HI, April 23–26, 2012, AIAA, Papers published on-line.

[71] AIAA/ASME/ASCE/AHS/ASC, 48th AIAA/ASME/ASCE/AHS/ASC Structures, Structural Dynamics, and Materials Conference 15th AIAA/-ASME/AHS Adaptive Structures Conference 9th AIAA Non-Deterministic Approaches Conference 8th AIAA Gossamer Spacecraft Forum 3rd AIAA Multidisciplinary Design Optimization Specialist Conference AIAA Dynamics Specialists Conference, Sheraton Waikiki, Oahu, 2255 Kalakaua Avenue, Honolulu, HI 96815, April 23–26, 2007, AIAA, Proceedings published on CD-ROM—AIAA Meeting Papers on Disc, 2007.

[72] AIAA/ASME/ASCE/AHS/ASC, The 44th AIAA/ASME/ASCE/AHS/ASC Structures, Structural Dynamics, and Materials Conference and AIAA/ASME Adaptive Structures Forum, Technical Papers, Norfolk, VA, April 7–10 2003, AIAA, Proceedings published on CD-ROM—AIAA Meeting Papers on Disc, April 2003.

[73] AIAA/ASME/ASCE/AHS/ASC, The 47th AIAA/ASME/ASCE/AHS/ASC Structures, Structural Dynamics, and Materials Conference, the 14th AIAA/ASME/AHS Adaptive Structures Conference, 8th AIAA Non-Deterministic Approaches Conference, 7th AIAA Gossamer Spacecraft Forum, and 3rd AIAA Multidisciplinary Design Optimization Specialist Conference and AIAA Dynamics Specialists Conference—Technical Papers, Newport, RI, May 1–4, 2006, AIAA, Proceedings published on CD-ROM —AIAA Meeting Papers on Disc, 2006.

[74] AIAA/ASME/ASCE/AHS/ASC, 52nd AIAA/ASME/ASCE/AHS/ASC Structures, Structural Dynamics, and Materials Conference 19th AIAA/-ASME/AHS Adaptive Structures Conference 13rd AIAA Non-Deterministic Approaches Conference 12nd AIAA Gossamer Spacecraft Forum 7th AIAA Multidisciplinary Design Optimization Specialist Conference AIAA Dynamics Specialists Conference, Sheraton Downtown, Denver, CO, April 4–7, 2011, AIAA, Papers published on-line.

[75] AIAA/ASME/ASCE/AHS/ASC, The 43rd AIAA/ASME/ASCE/AHS/ASC Structures, Structural Dynamics, and Materials Conference and AIAA/ASME Adaptive Structures Forum, Technical Papers, Denver, CO, April 22–25 2002, AIAA, Proceedings published on CD-ROM— AIAA Meeting Papers on Disc, April 2002.

[76] AIAA, SciTech 2015: The 2nd AIAA Science and Technology Forum and Exposition, Gaylord Palms Hotel & Convention Center, Kissimmee, FL, Jan. 5–9, 2015, AIAA, Papers published on-line.

[77] AIAA, 48th Aerospace Sciences Meeting Including the New Horizons Forum and Aerospace Exposition, Orlando, Florida, Jan. 4–7, 2010, AIAA, DOI 10.2514/MASM10.

CHAPTER 4

Deployment Motions and Forces

Joe Footdale
LoadPath, Albuquerque, NM, USA

Ed Boesiger
Lockheed Martin, Sunnyvale, CA, USA

Satoru Ozawa
JAXA, Tsukuba, Japan

Hiraku Sakamoto
Tokyo Inst. Tech., Tokyo, Japan

Yoji Shirasawa
JAXA, Tsukuba, Japan

Sungeun Jeon
LoadPath, Albuquerque, NM, USA

Mathieu Robroek
Airbus Defence and Space Netherlands B.V, Leiden, Netherlands

4.1 INTRODUCTION

Deployable space structures in general are considered high risk and, in most cases, are mission critical. Standard practices have been implemented to mitigate those risks through careful mechanical design [1] and thorough test programs. For example, NASA General Environmental Verification Standard (GEVS) NASA-GSFC-STD-7000A [2] lists the justifications and required tests for spacecraft with deployable systems. By definition, the structural responses of LULS are heavily influenced by gravitational loads and require compensation systems to simulate the microgravity environment during ground testing. This increased sensitivity to the test environment interactions requires additional considerations in the analysis and testing of the structure deployment to satisfy the flight hardware verification standards. This chapter presents the methods, procedures, and best practices for characterizing the LULS structural response *during* the deployment process. Deployment repeatability and deployed structural testing techniques are covered in Chapter 5 and Chapter 6, respectively. Many of the underlying technologies presented herein are discussed in other chapters in the book. For example, the various methods of gravity compensation are presented in Chapter 3. This chapter presents a brief review/description of the relevant technologies, but emphasis is placed on what should be selected and the reasoning why.

In addition to the categories introduced in Chapter 2, deployable structures can be further categorized according to the rate of deployment. Design standards [1] recommend deployments be controlled or restrained, where practical, for the full range of motion. The deployable structure should not be designed to rely upon inertia for successful deployment. However, advancing the state-of-the-art of LULS has led to the development of another category, which involves structures where deployments are not quasi-static. They are commonly referred as "free" deployables. These structures often use stored strain energy from compliant mechanisms to provide the deployment motive forces and do not incorporate rate control mechanisms. Examples include the MARSIS boom [3] and the Springback parabolic reflector [4]. This distinction between quasi-static and free deployment is important because it heavily influences the design of the deployment test and ground support equipment. These "fast" deploying structures eliminate several gravity compensation system architectures (helium balloons are not an effective selection, for example) and might require additional effort to assess the impact on the spacecraft. Additional deployment testing challenges incurred with fast deployments include parasitic loading from inertial effects of the offload equipment, air drag, and requiring a metrology system capable of high acquisition rates.

The influence of the test environment on the structure during deployment testing can result in significant deviations in the observed kinematics and loads when compared to the on-orbit configuration, as supported by the examples above. Uncertainty in the deployment kinematic model is minimized by verifying it with ground test data. Section 4.2 reviews the main objectives and requirements for the ground deployment test program aimed to minimize this uncertainty and mission risks.

Deployable structures designed for the low disturbance loads in the microgravity environment are extremely sensitive to the ground support equipment needed to perform ground deployment tests. The interactions between the structure and ground support equipment must be included in the model for proper correlation and model updating procedures to be performed. Although the main intent of this book and chapter is to present best practices in conducting structural tests, the unique considerations and methods for modeling the deployment are presented in Section 4.2. Other general modeling best practices are presented to assist in programs where system-level deployment testing is not possible.

Section 4.3 describes the methods and best practices of conducting ground deployment tests. This largely comes down to minimizing the influence of the gravity compensation and other test equipment needed. Common transducers and metrology systems are also presented.

Section 4.4 describes the test programs of a few deployable structures in detail.

4.2 DEPLOYMENT TESTING OBJECTIVES

The primary objectives for deployment testing, which are largely derived and discussed in Section 2.4.5.2 of the GEVS standard, are summarized below. These objectives are not unique to NASA spacecraft and can be found in European Space Agency (ESA) and Japan Aerospace Exploration Agency (JAXA) Engineering Requirement Guideline (JERG) standards [5] as well.

4.2.1 VERIFICATION OF DEPLOYMENT KINEMATIC MODELS

GEVS 2.5.4.2.a.—Analysis states [2]:

> *. . . a kinematic analysis of all deployment and retraction sequences shall be conducted to ensure that each mechanism has adequate torque margin under worst-case friction conditions and is capable of withstanding the worst-case loads that may be encountered during unlatching, deployment, retraction, relatching, or ejection sequences. In addition, the analysis shall verify that sufficient clearance exists during the motion of the mechanisms to avoid any interference.*

Although this requirement in the standard is with respect to the kinematic model, it is listed here as a test objective because deployment testing of the structure is desired to correlate with the model and reduce the uncertainty. The term *desired* was used as opposed to *required* because full-scale deployment testing at the system level may not be feasible. In these circumstances, deployment testing at the highest possible subsystem or component level should be performed. Large ultra-lightweight structures are inherently highly nonlinear systems with respect to the deployment mechanics. A number of flight programs have shown that, even with exhaustive subsystem test programs, variations from the kinematic model and flight deployment data were significant. The deployable MARSIS lenticular jointed-boom antenna [3] initially failed to fully deploy due to hinge stalling at one location. A deployment test of the complete boom was determined to be not feasible by the development team because a gravity compensation system that did not overly influence the deployment motions could not be integrated. The program instead relied upon component-level testing of the boom hinges with Monte-Carlo simulations to analytically predict the range of possible boom kinematic paths and deployment times. During the anomaly resolution effort, the results from an updated model showed that the hinges experienced more extensive back-buckling than those observed in the original analysis. An extensive testing and analysis effort was undertaken over the subsequent months to better characterize the boom. The updated model showed excellent agreement with flight dynamics data once the boom fully deployed. Further examples can be found in Refs. [3, 6, 7].

4.2.2 TEST LEVELS AND MARGINS

GEVS 2.4.5.2.c.(2)—Subsystem Testing states [2]:

> *For each mechanical operation, such as appendage deployment, tests at nominal-, low-, and high-energy levels shall be performed. One test shall be conducted at the most probable level that will occur during a normal mission (the nominal level). The test will establish that functioning is proper for nominal operating conditions and baseline measurements will be obtained for subsequent tests...*
>
> *If a margin test cannot be conducted at the subsystem level due to its size and complexity these verification tests shall be performed at the highest level of assembly possible and the results combined to provide subsystem performance.*

This section gives the justification and motivation to perform system-level deployment tests for the structure subjected to the extreme environments that create the low- and high-energy levels. The desired output of this testing is to determine the overall torque margin, which is defined in more detail in GEVS Section 2.4.5.3. A poorly designed or absent test can yield significant errors in the observed maximum torque required to deploy the structure. If the test support is inadvertently assisting the deployment, the actual torque margin will be under-predicted, and insufficient torque can result in failure to deploy, as shown in the MARSIS example. The opposite case, where the excessive torque margin is available, is also possible. This typically results in an over-designed structure, which is not mass optimized, but it could also lead to high deployment rates that result in the high end of deployment shock with possible damage to components.

4.2.3 END OF DEPLOYMENT SHOCK

GEVS 2.4.4.1 [2] states:

> *Also the impact on deployable devices as they reach their operational position at the "end of travel" is a likely source of significant shock. When hardware contains such devices, it shall be exposed to each shock source twice to account for the scatter associated with the actuation of the same device.*

The shock loading is important to the spacecraft attitude control system and for its potential to damage the structure. These effects are more common in free deployable architectures. Shock loads are also common in structures with high strain composite (HSC) components that create impulse loading during snap-through events at end of travel even if the deployment kinematics are controlled.

4.2.4 DEPLOYMENT REPEATABILITY

Deployable structures typically have an end of deployment position tolerance requirement that they must satisfy. Deployment testing is performed multiple times to assess the repeatability of the final deployed position or shape. The test methodologies and best practices for this aspect of deployable structure testing are presented in Chapter 5.

4.3 DEPLOYMENT ANALYSIS

One of the main reasons for conducting deployment test is to verify and validate the analytical model. Doing so decreases uncertainty in the on-orbit deployment predictions. The influence of the test environment (including, but not limited to, gravity) on the structure can alter the deployment compared to the on-orbit configuration. This section

- Identifies typical factors and their importance to the model.
- Summarizes the general methods and emphasizes the unique considerations for modeling the deployment.
- Lists the requirements for model verification/validation when a full-scale system test cannot be performed. It discusses the criteria that need to be met for the program manager to accept that the structure is qualified for flight.
- Describes two examples of actual LULS missions in Japan: the large deployable reflector and the solar sail demonstrator. During these missions, a full deployment test was not practical.

4.3.1 VERIFICATION STRATEGY WITH ANALYSIS AND TEST

In the early design phase of a LULS, many analyses are generally conducted to ensure the system can satisfy its mission requirements. Static and dynamic structure analyses are used also to simulate the behavior of the deployment system.

Many factors can affect the behavior of the LULS structure because it has high flexibility and can be deformed with low forces or torques that are normally negligible. For example, buckling and plastic deformation of a structure that implements cables or membranes can crease, wrinkle, or slacken, which can affect the kinematics and dynamics of the deployment system. These factors that should be considered vary from mission to mission, so there is no commonly used analysis method for which the validity has been verified sufficiently.

For these reasons, it is recommended that the validity of the analysis model be verified with an element examination, where a component- or subsystem-level test is performed specifically for model validation. In the case of large deployable reflectors of communications satellites [8, 9] or the IKAROS solar sail demonstrator program [7, 10], many element examinations were conducted to evaluate the effect of structural parameters such as damping coefficient, bending stiffness, and crease stiffness. These results were reflected to the analytical model. Simulation results from the model were also compared with the results of element examination to revise the model. The analysis model constructed in this way in the design phase can be used for the verification of the system feasibility as well as for the determination of the test conditions in the development phase and the operating rules at the operation phase.

4.3.2 IDENTIFICATION OF VERIFICATION ITEMS

In the design phase, system requirements of the deployment system will be determined to satisfy the LULS mission requirement. Some of the system requirements are difficult to verify by tests in the development phase, so these should be transformed to items that can be controlled and verified by the test. This transformation is also examined by numerical simulation and sensitivity analysis using analytical models. Through these processes, it is important to identify the factors that prevent achievement of the mission and system requirements.

Typical examples of system requirements and control items in LULS deployment system are listed as follows:

- System requirements for the deployment process:
  - Allowable variation in attitude of the spacecraft during deployment operation.
  - Allowable vibration, shock, and deformation of the deployable structure during the deployment process.
  - Deployment duration.
- Typical control items:
  - Mass property of the spacecraft.
  - Mass distribution of the deployable structure.
  - Variation of the operating time of actuators in the release mechanism.
  - Variation of the speed of the motor for expanding the deployable structure.
  - Variation in material density and thickness of thin, composite materials.
- System requirements for after completion of the deployment:
  - Surface accuracy.
  - Thermal stability.
  - Dimensional stability.
- Items that the deployment achievement rate can be controlled by:
  - Structural, mechanical, and material properties of the deployable structure.
  - Linear coefficient of expansion of each structure.
  - Thermal and optical property, charging characteristic.

Note that this list contains control items that will cause a low disturbance force, such as solar pressure and electrostatic force. These are normally ignored for other types of spacecraft structures; however, even low forces and torques can deform LULS structures, making these sources of load important.

4.3.3 ASSIGNMENT OF VERIFICATION METHODS

The Finite Element Model (FEM) method is commonly used for spacecraft structure analysis. For lightweight structures, membrane models or shell models are used as the mathematical model; however, some additional issues should be considered particularly for LULS structures.

LULS structures can be deformed globally with little force, so they can take quite different forms at slightly different energy levels. Furthermore, structures that contain membrane or cable components may be subject to drastic changes in stiffness in each time step due to wrinkles or slack, so their equations of motion will be stiff differential equations. For these reasons, several models and numerical methods have been proposed to enhance the convergence performance and stability of the numerical analysis with a limited computational resource.

To simulate the behavior of the membrane and cable structure with a simple method, the Multi Particle Model (MPM) method was developed [10, 11]. In this model, each element of the structure is substituted by particles connected by springs and dampers. This leads to the reduction of the computational cost at the sacrifice of the precise prediction of the internal stress condition. This modeling method, which has been validated by comparison with FEM simulation in a typical case, has significant computation advantages in simulating dynamic deployment motion in which the inertial force is dominant over the internal tensile forces. On the other hand, for simulations where the internal tensile force is dominant, such as the prediction of the shape of the structure after the completion of the deployment, FEM modeling should be used.

4.3.4 LARGE DEPLOYABLE REFLECTOR

The large deployable reflector (LDR) on the Engineering Test Satellite VIII (ETS VIII), shown in Fig. 4.1, is a modular mesh antenna that consists of 14 modules and has been verified through a series of ground tests and numerical analyses before the launch [8].

On ground tests of large deployable structures, it is extremely hard to measure precise deployment forces even with gravity compensations. The LDR is gravity-compensated with hanging cables in the test setup. The springs installed in the LDR are the source of the deployment energy and expand the reflector to the fully deployed state. The deployment torque is specified as a relatively small value to deploy and not to destroy the lightweight reflector whereas the gravity torque goes much greater than the deployment force as the LDR becomes larger.

During the development of the LDR, Meguro, Ishikawa, and Tsujihata [8] devised an index that quantified the

Fig. 4.1 Ground test setup of a large deployable reflector. (Credit: JAXA.)

difficulty of performing ground deployment tests. The Difficulty of Deployment Test (DDT) index was defined as the ratio of gravity torque to deployment torque required for micro-gravity deployments. This index was related to measured estimation errors of the deployment force. An estimation error of 10% in the deployment force was found for a structure with a difficulty index = 1. The study concluded that ground deployment testing for structures with an index = 1000 was not practical. Examples of DDT value for multiple large space structures are given in Fig. 4.2. The DDT was used to identify the LDR as a deployable structure whose full deployment on-orbit is impossible to guarantee through ground testing alone. Therefore, the LDR is divided into small modules that can be verified by ground tests. The full deployment of LDR is guaranteed by a combination of the module ground tests and numerical analyses.

Figure 4.3a shows a picture of a seven-module ground test, and Fig. 4.3b is a comparison of the ground test result and the numerical analysis. In the numerical analysis, the analysis model includes the test setup and gravity compensation devices aside from the reflector. The comparison of ground tests and numerical analyses has been done with 1-, 3-, 7-, and 14-module reflectors. On top of ground tests, LDREX-2, the in-orbit test, was carried out. Figure 4.4a is an artist's view of the LDREX-2. Figure 4.4b is a comparison of test result and numerical analysis [12]. Through the series of comparisons, the numerical

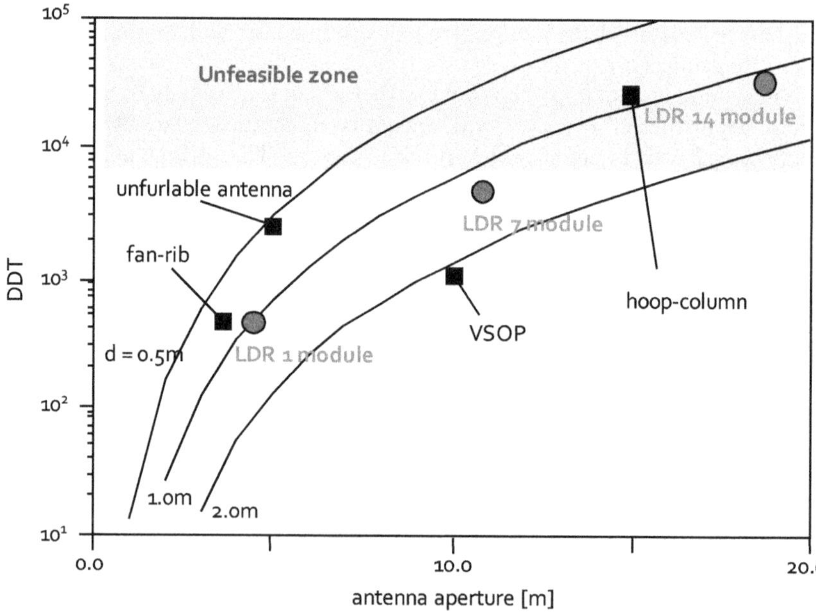

Fig. 4.2 DDT factor for several large deployable structures [8].

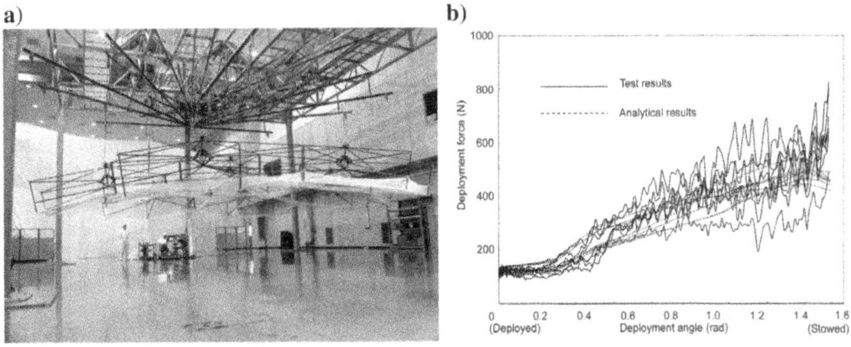

Fig. 4.3 Comparison of ground test and numerical analysis. (a) Picture of ground test. [Credit: JAXA.] (b) Comparison of deployment forces of ground test and numerical analysis [12].

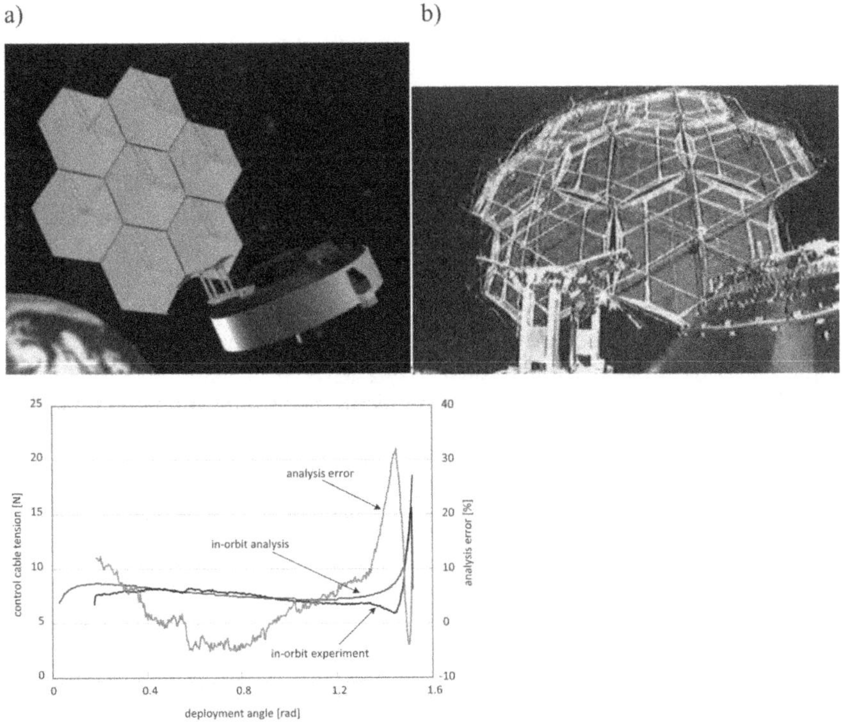

Fig. 4.4 Comparison of in-orbit test and numerical analysis. (a) Concept illustration of LDR in-orbit experiment (LDREX-2). (Credit: JAXA.) (b) Comparison of deployment forces of in-orbit test and numerical analysis [2].

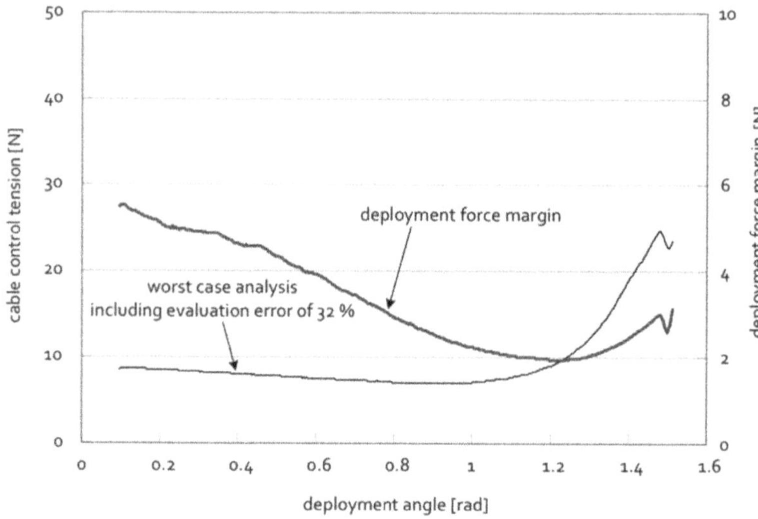

Fig. 4.5 Deployment analysis of LDR [12].

model of LDR has been updated and verified. Finally, the full deployment analysis was performed with the updated numerical model. The deployment analysis result in Fig. 4.5 shows that the deployment margin is greater than two and guarantees the full deployment in orbit. All the deployment analysis was performed by an analysis method that combines FEM and multibody dynamics [13].

As with deployment, the surface accuracy is guaranteed by numerical analyses. The worst possible surface error is calculated by summation of each module surface error. The module surface error was measured in ground tests. Figure 4.6 is an evaluation of surface error [14].

The research project [9] for a communications satellite with a larger antenna aperture (shown in Fig. 4.7) after ETS VIII has employed the same verification methodology, which is a combination of analyses and tests.

4.3.5 SOLAR SAIL DEMONSTRATOR

The solar sail demonstrator IKAROS successfully deployed its 14 m × 14 m sail in 2010 [15, 16, 17]. Figure 4.8 shows the spacecraft. For the deployment of the sail, a two-stage sequence was

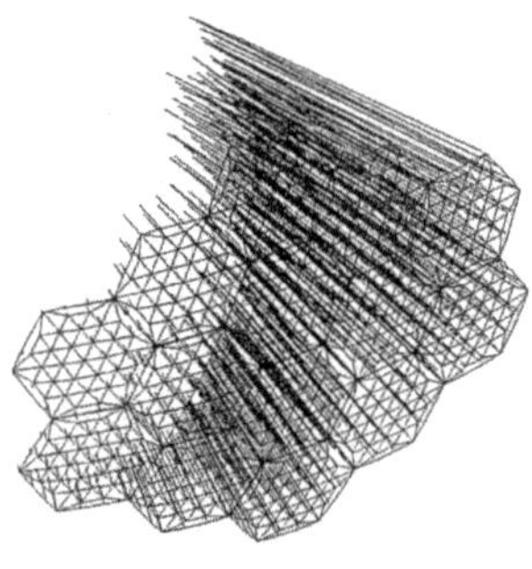

Fig. 4.6 Surface error analysis of LDR. The lines projected from the reflector surface indicate surface error relative to the intended parabolic shape [14].

Fig. 4.7 Deployment test of LDR for future communications satellites [9].

adopted, as shown in Fig. 4.9. The deployment sequence consists of the first stage, quasi-static deployment, and the second stage, dynamic deployment. For reliable deployment, two conditions were investigated. In the first stage, dynamic effects should be excluded as much as possible in the deployment behavior. The deploying sail should maintain the balance with centrifugal force. In the second stage, the membrane folded crosswise is developed dynamically. At this time, the deployment behavior depends on many structural characteristics that are hard to identify, such as the damping coefficient of the membrane and high-frequency vibration characteristics. This uncertain system should complete the deployment smoothly without falling into undesired stability modes. In the IKAROS Project, numerical analyses were used intensively in this verification process. This section describes how the analyses were used in the project. Two kinds of numerical analysis techniques—MPM and geometrically nonlinear FEM—were used for the design exploration to satisfy the aforementioned two conditions, and for the

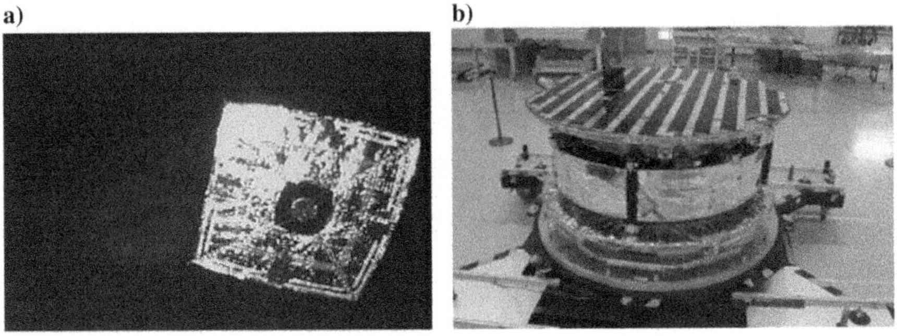

Fig. 4.8 Solar sail demonstration spacecraft, IKAROS. (a) Picture taken by deployable camera. (b) Flight model in final assembly phase [17].

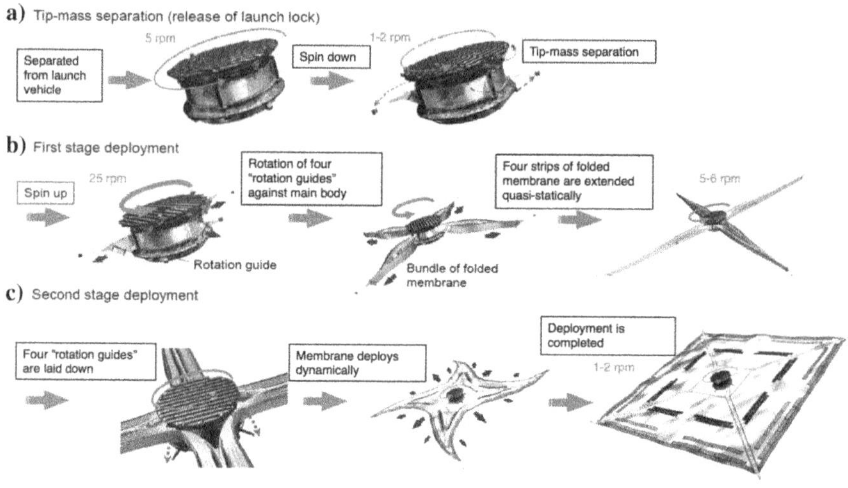

Fig. 4.9 Solar sail IKAROS's deployment sequence [18].

evaluation of the robustness against parameter uncertainty. These two kinds of models were used as follows.

First, MPM expresses membranes and tethers as the network of concentrated masses connected with springs and dampers. In IKAROS, the models for the first stage deployment and second stage deployment were developed, as shown in Fig. 4.10 [10, 11, 19]. From the stowed configuration, the transient response was calculated by explicit time integration. The accuracy of MPM models was evaluated by comparison with balloon test and sounding rocket test results [20, 21, 22]. The advantages of MPM models are that the model is simple and computation time is relatively short. Therefore, many analyses with combinations of various design parameters were possible. As a result, large parametric spaces were explored, and selected parameters could be evaluated. However, the stress state in membranes is not modeled rigorously in MPM. Therefore, although the global phenomenon

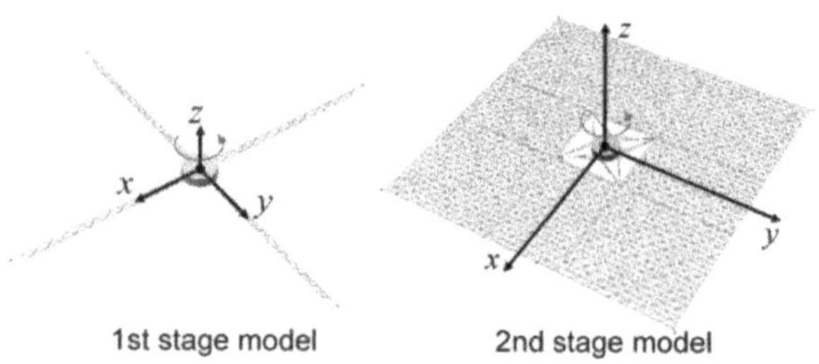

Fig. 4.10 MPM models for IKAROS [10]. (Credit: JAXA.)

caused by inertial forces can be estimated, MPM is not good at searching for undesired stability modes, which are mainly caused by the elastic forces.

Therefore, in the IKAROS Project, the accuracy of MPM models was evaluated by comparing some typical MPM simulation cases with FEM models. FEM models were constructed using in-house code for first stage deployment and second stage deployment [23, 24]. In the calculations, an implicit time integration was used where the force balance should be converged in each time step. In the second stage development, the membrane was modeled using a special plane stress element that takes wrinkling and slackening of the membrane into consideration [18]. An example of analysis results of the second stage deployment is shown in Fig. 4.11. The FEM model rigorously represents the stress distribution in the membrane compared to the MPM model. Although a relatively long computation time is required, FEM was very useful to evaluate the accuracy of MPM analyses, confirming that no undesired stability mode exists during the deployment.

When the flight data of IKAROS were observed, two phenomena were not predicted during the pre-flight analyses. First, the actual sail deployment was asynchronous. Figure 4.12 shows the images taken by the four monitor cameras onboard the bus. The shape after the first stage deployment was not symmetrical, and in the second stage deployment, one side of the sail was deployed with about a 30-second delay. As shown in Fig. 4.13, a MPM analysis agrees with the flight data well when this delay of one side sail was considered [10]. The cause of this deployment delay has not been fully understood yet. Variations in stiffness of the fold lines of the membrane were suspected as a possible reason of asynchronous deployment. This is because asynchronous deployments due to crease stiffness were observed in a ground deployment experiment using a small membrane model of 455 mm × 455 mm, as shown in Fig. 4.14. However, the dimensionless analysis considering the centrifugal force and the elastic force in the membrane

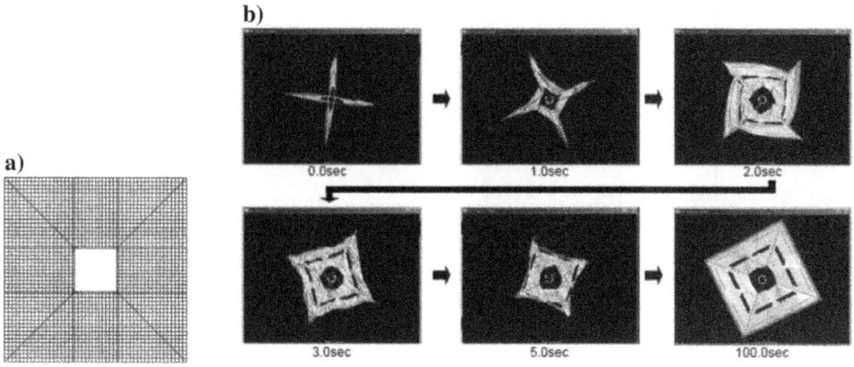

Fig. 4.11 Pre-flight deployment analysis of second stage deployment using FEM [18]. (a) Finite-element mesh. (b) Deformed shape.

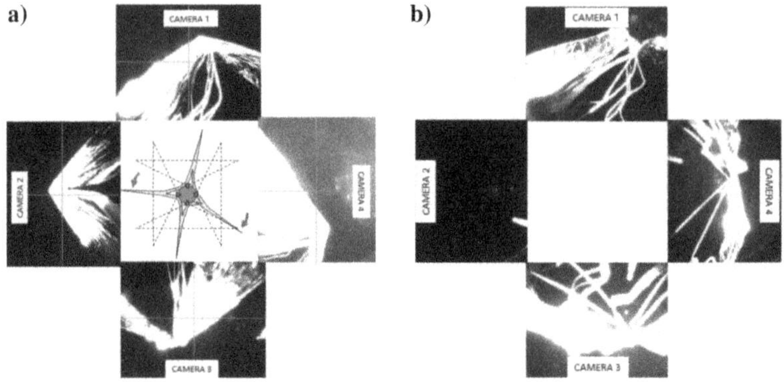

Fig. 4.12 Side view monitor camera images on IKAROS. (a) After first stage deployment. (b) At 27 sec, after the start of second stage deployment [25].

shows that the centrifugal force is dominant at the scale of IKAROS, and the behavior seen in the small model hardly occurs at the size of IKAROS [25]. In the stowed configuration, IKAROS had thin membrane strips to hold the winded membranes. These membrane strips may have prevented the deployment of one side of the sail.

The second phenomenon that was not considered in the pre-flight analysis was the non-flat equilibrium shape of the sail. After the deployment of the

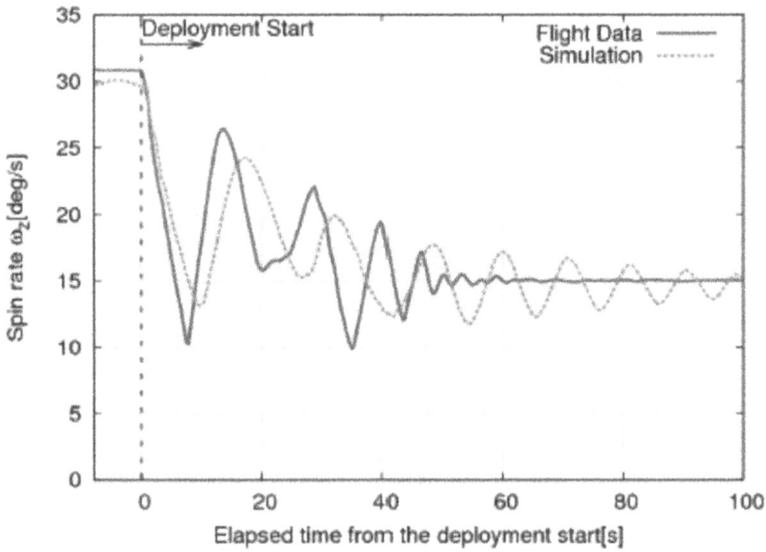

Fig. 4.13 Comparison of spin rate of MPM simulation result and flight data for second stage deployment [10].

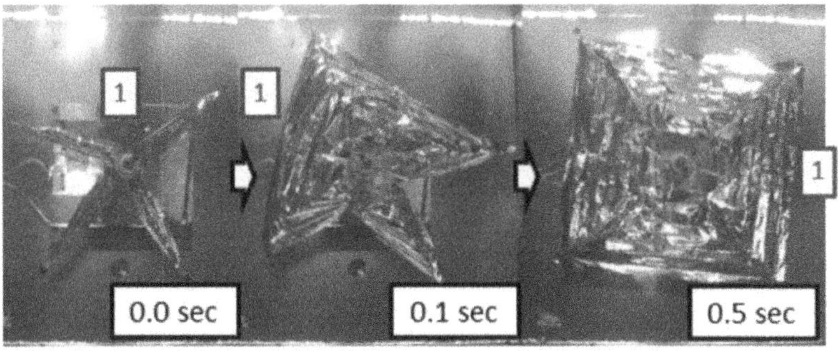

Fig. 4.14 Non-synchronous deployment behavior observed in a vacuum chamber using 455 mm × 455 mm square membrane [25].

IKAROS sail in orbit, two deployment cameras, separated from the spacecraft bus, took pictures of the sail, some of which are shown in Fig. 4.15. Then the sail shape was estimated by image processing [15] to find that the sail deflects about 500 mm toward the sun (see Fig. 4.16), which was not initially expected. Therefore, after the flight, a new kind of numerical analysis model was constructed to reproduce this non-flat sail shape. Geometrically nonlinear FEM is used, but the membrane is now modeled by shell elements, which considers the small bending stiffness of the membrane. The shape of the quarter part of the sail was calculated [26]. One of results is shown in Fig. 4.17. The result shows that the deformation of thin-film devices, such as a thin-film solar cell attached on the membrane, causes the

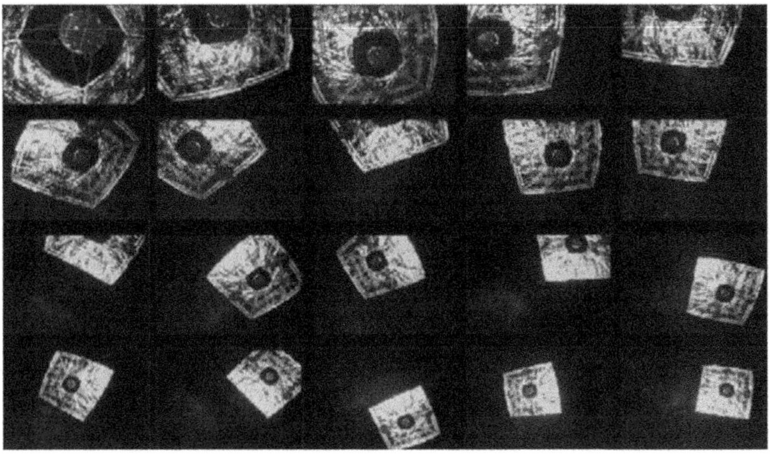

Fig. 4.15 Series of images taken by deployable camera onboard IKAROS [15]. (Credit: AIAA.)

Fig. 4.16 Sail shape estimated by deployment camera images. Maximum deflection is estimated at approximately 500 mm [15]. (Credit: AIAA.)

deformation of the entire membrane. Consequently, the sail becomes non-flat, and the out-of-plane stiffness is significantly increased.

As described above, the IKAROS Project intensively used numerical analysis models to evaluate the design before launch, and some new phenomena that were not fully considered before the launch were found on orbit. Therefore, for successful missions, it is important to

- Accumulate many flight experiences to improve analysis models so that models can consider new phenomena.
- Carry out a large number of analyses to evaluate the robustness of the design against all possible uncertainties in the model.

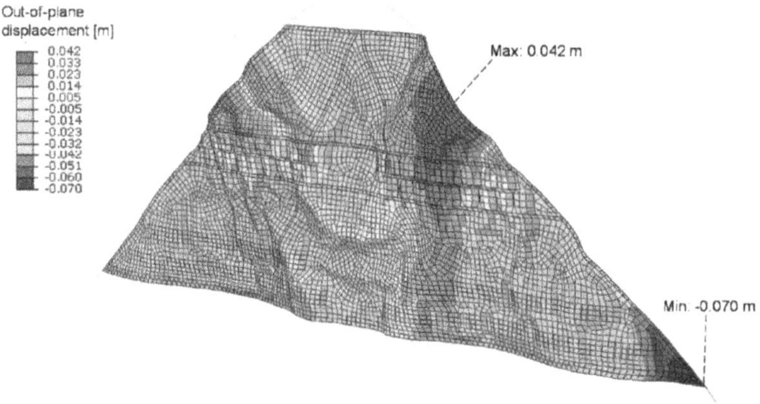

Fig. 4.17 Sail shape prediction calculated by FEM with shell elements [26]. (Credit: AIAA.)

4.4 DEPLOYMENT TESTING

Chapter 2 of this book includes the fundamental discussion of the development of overall test programs for various spacecraft classes and associated requirements. This section expands on that discussion as it relates to performing deployment tests.

The overall model uncertainty is minimized with deployment testing performed in a like environment. The deployment test is designed to mimic the on-orbit conditions subject to feasibility, acceptable risk, cost, and schedule constraints of the program. Any deviations from the expected environment applied to the structure during testing must be well understood and predictable. Any test-induced loading must also be included in the model for verification and validation procedures. These factors are then removed from the model to predict the on-orbit deployment. If a deployment test is poorly designed or is not feasible without causing excessive deviations to the natural kinematics, an inaccurate model and/or deployment failures can result from inadequate torque margins or anomalous deployment motions.

Another approach to reducing the model uncertainty is to design the structure to be insensitive to gravitational and other test-induced loads. Based on the calculated expected loads to be imparted on the structure from the ground test equipment and deployment environment, factors such as deployment torque and structural stiffness or strength can be increased to keep the parasitic test loads within a small fraction of the structure deployment torque. As mentioned before, this results in an over-designed structure for the actual mission requirements.

4.4.1 DEVELOPING THE TEST PLAN

The following sections discuss some key test plan parameters that are significant to LULS deployment testing. Refer to Chapter 2 for a comprehensive discussion of test plan development.

4.4.1.1 NUMBER OF DEPLOYMENTS

For new architectures, multiple deployment tests should be performed on flight-like engineering units to demonstrate repeatable and reliable results. A minimum of three deployments in similar test conditions should be performed to develop a data trend and to generate some statistical significance. The ultimate number of tests and the environment in which the tests are performed is again subject to feasibility, acceptable risk, cost, and schedule constraints of the program.

4.4.1.2 BOUNDARY CONDITION MODIFICATIONS

In addition to designing and conducting deployment tests in a simulated "like" environment, the deployment test can be performed with intentional modifications to the structure or test setup. The objective of this modified test is to generate additional, linearly independent data to minimize the uncertainty in the deployment model. The modifications should be easy to model such that the structural contributions do not increase the number of fit parameters. Examples of boundary condition modification are varying the number of gravity offload connection points, modifying the structure inertial properties with added point masses, and performing the test in multiple gravity orientations.

For example, the Advanced Synthetic Aperture Radar (ASAR) on the European ENVISAT spacecraft performed deployment testing in two gravity orientations using a different gravity compensation system for each configuration [27]. The antenna structure consisted of five flat panels connected by revolute hinges, where all rotation axis vectors were parallel. The first deployment test was conducted with the gravity vector normal to the deployed panel surface. The panels were suspended by a system of cables with counterbalance weights to compensate for the panel masses. A second deployment test was then performed with the gravity vector parallel to the hinge axes. The panels were supported by air bearings that were free to traverse on a flat floor.

4.4.1.3 STRUCTURES WITH HSC COMPONENTS

HSCs encompass structures that undergo large deformations through composite materials capable of large strains [28]. This capability is utilized in advanced lightweight deployable structures to avoid the complexity, weight, and cost limitations of the mechanical approach by using material deformations (such as material stretching) to fold a structure. This section summarizes notable considerations for deployment testing of LULS structures with HSC that should be included in the test plan. All of these factors impact the available deployment torque that must be characterized to ensure sufficient margins exist.

- *Humidity/moisture.* Humidity levels within the lab environment are typically controlled to mitigate electrostatic discharge (ESD) events. Humidity levels can also heavily influence the deployment kinematics of structures, particularly those incorporating HSC components, where the mechanical properties of the resin system used can vary. The Roll-Out De-Orbiting devices (RODEO) [29] consisted of lightweight film attached to a simple, ultra-lightweight, roll-out composite boom structure. The structure was launched to low Earth orbit (LEO) for a flight experiment and showed an eventually successful, but anomalous, deployment where the boom motion was intermittent and not constant as observed in ground testing. This anomalous deployment was shown to be caused by moisture saturation in the composite boom acquired from the relatively high humidity environment of the eastern United States, where the boom was located for integration and launch. Additional laboratory testing determined that an induced high-humidity environment caused the composite to absorb moisture, plasticizing the resin and significantly reducing the stiffness and stored strain energy of the boom. The effect was duplicated in laboratory conditions and is completely reversible. The boom would deploy as expected after extended exposure to a relatively dry environment.

- *Stow duration*. When HSCs are used, they are often subjected to prescribed loads or enforced displacements for long periods of time. An example of this is a compliant composite flexure in a deployable structure (such as MARSIS [3]) that may be held in a folded, or stowed, configuration for many months between assembly and ultimate deployment. Over these timescales, it has been observed that many fiber/resin systems exhibit time-dependent effects that usually, although not always, correspond to a degradation of behavior in comparison to predictions made assuming elastic properties or short-term experimental data. The prediction and understanding of time-dependent properties is clearly vital if HSCs are to be used with confidence in design.
- *Temperature*. Environmental temperature has similar effects as the stow duration described above. The two are directly related. Stow duration effects can be accelerated by subjecting the HSC to elevated temperature when in the stowed/stressed state.

As described in the "Deployment Testing Objectives" section, earlier in this chapter, deployment tests should be performed at normal-, low-, and high-energy states. The structure's stow durations and laboratory environmental conditions should be designed to achieve the span of energy states to properly assess the deployment torque margins and possible range of deployment kinematics.

4.4.2 TEST PLANS WHEN FULL SYSTEM TEST IS NOT PRACTICAL

Full-scale, system-level deployment testing may not be feasible or practical for several reasons. Structures, such as the deployable solar array wings now aboard the International Space Station, can exceed vacuum chamber dimensions. Test programs may also not be budgeted to test in a large vacuum chamber. It was not possible to do ground deployment testing of the 20 m diagonal IKAROS solar sail, which was spin-deploy/stabilized [30].

As previously described in Section 4.2, earlier in this chapter, the DDT index was developed to function as a useful tool guide the decision to perform substructure or subsystem tests in place of a full system test. For the typical LULS structure, large, complicated gravity compensation arrangements are needed to perform ground deployment tests. In these cases, the characteristics of the ground test system are evaluated rather than the characteristics of the structure. Therefore, the accuracy of ground deployment testing must be examined. If the estimation error is several tens of percent of the designed deployment force, more force margin must be added to the nominal design values. If the error is several hundreds of percent of designed deployment force, the test results cannot be readily trusted.

For structures where full-scale qualification deployment testing is not feasible, two possible solutions are to test on a scaled structure and to test all subsystems independently.

4.4.2.1 SCALED STRUCTURES

For the IKAROS solar sail program, centrifugal testing was performed on a 0.63 m diagonal development unit. Data from the test was used to correlate with the analytical model, where the approach was used to predict the full-scale deployment and deployed structural responses [19]. The data sent from IKAROS in space suggested an important phenomenon for the deformation properties. The deflection of the sail induced by solar pressure was smaller than that predicted preflight, which suggests that the out-of-plane stiffness of the sail was higher than expected. Further analyses concluded the high out-of-plane stiffness of the IKAROS membrane can be induced by the increased bending stiffness and by the changed deformation mode due to the curvature of the solar cells that were populated on the sail membrane surface [31].

A scaled model of the basic module of the LDR of the ETS VIII was tested in a micro-gravity environment from the parabolic flight path of a specially equipped jet [8]. The estimation error, which was found to be about 10 percent, was due to the approximations used to model non-structural mass elements (such as mesh and cables). For the full-scale reflector, whose size is 19 m $\times$ 17 m, it was virtually impossible to evaluate its actual characteristics. Testing on the scaled model would not provide sufficiently accurate data for flight qualification.

Deployment testing on scaled structures should, where possible, be reserved for developmental efforts in determining the overall feasibility of the architecture. The potential missions for a solar sail structure [32], would require deployable booms on the order of 80–160 m in length. Development efforts aimed to achieve the Technology Readiness Level of 6 requires testing of a prototype unit in a like environment, which translates to deployment testing in a vacuum. Even with the largest currently available test facility [33], this limits the individual boom length to approximately 30 m. To maximize the usefulness of the test program, the length of booms designed for the larger application was truncated as opposed to scaled in this example. Truncation was deemed to give more accurate representation of the full-scale system as compared to dimension scaling, which would require optimizing the boom design for the shorter length that could be tested in a thermal-vacuum environment.

4.4.2.2 SUBSYSTEM TEST ONLY

Common test practice focuses on component testing for items that are vacuum sensitive or need definitive measurements, such as deployment torques at thermal vacuum for a hinge mechanism. These component- or subsystem-level results are included in the system-level model to predict the structure deployment. For subsystems that are highly linear, the principle of superposition effectively predicts the coupling of the multiple systems. Large uncertainties can arise when geometric or material nonlinearities cannot be neglected, such as is the case for structures implementing HSC components or applications that require high-precision position and structural stability. The scope and design of the subsystem testing should be such that it spans the full range of expected states to ensure the component model will accurately predict all potential structural states. The underlying issue with subsystem testing with nonlinear components is that dominant structural modes in the fully coupled system may not be observed in the subsystem test. Therefore, the full range of expected states to apply in a subsystem test is not readily achieved.

Real-time hybrid testing [34] is an active research field that addresses this nonlinearity issue when synthesizing component models by creating the *in situ* environment for component or subsystem testing. This testing methodology is also referred to in the literature as Dynamically Substructured Systems (DSS) [35]. The basic concept of the hybrid test methodology lies in the decomposition of a structure into two or more substructures, where one substructure is a physical component of the remaining structure to be simulated numerically. During the test, measurements are taken at the location where the component physically interfaces with the remaining structure. These measurements are the inputs to the model of the coupled structure. The resulting reactions calculated from the model are then applied to the test component using an actuation system. Conducting this process in effectively real time virtually replicates a full system test. This method of testing requires specialized test equipment and control algorithms. The cost, complexity, and level of uncertainty with this relatively new methodology have limited its use in LULS applications to date. With continual advancements in the process, it may become an effective tool to decrease uncertainty for large structures where full system-level testing is not practical or feasible.

4.4.3 GRAVITY OFFLOAD/NEGATING

Chapter 3 discusses the various technologies and methods for negating the effects of gravity and how to properly design such a system. This section provides guidance about which technology or method to select, and why, for different classes of deployable structures. It is desired to keep the parasitic loads imparted by the offload system to a small fraction of the nominal offload force (within $+/-5$ percent is a typical goal). Keeping within this tolerance should prevent the offloading equipment from significantly influencing the structural response such that it does not give a false positive to the test nor cause a false negative failure that costs time and money to resolve. Variations in the offload force should be applied to the deployment model to determine the response sensitivity, which gives insight to the required complexity of the offload system architecture.

Orientation of the structure with respect to gravity is an important factor in the stabilization effect it has on the deployment. ATK's Coilable SAILMAST [36] truss is designed to deploy linearly along the axis defined by the boom deployed length. The deployment is performed perpendicular to the gravitational field to minimize that stabilization and allows the micro-gravity environment to be simulated to get accurate torque margin measurements.

Structure orientation is not obvious for deployments in more than one dimension, such as the Composite Deployable Telescope (CDT) [37], developed by the Air Force Research Laboratory (AFRL). The motion profiles for the offload connection points, shown in Fig. 4.18, were calculated for two different gravity orientations. The added challenges of negating the gravitational loads for this structure were the fast deployment rate and the extreme lightweight properties. The configuration selected had the structure deploying such that the membrane plane was perpendicular to the gravity field. This configuration minimized the deviation of applied offload force over the range of motion when a series of soft extensional springs was used. The relatively large height of the offload cable attachment points resulted in a negligible parasitic lateral load of approximately 3 percent of the vertical offload force.

The compensation system must not artificially control the deployment path. Revisiting the deployable truss, an air bearing system rigidly connected to the truss, would not be appropriate because it would impose a vertical displacement constraint for a truss that has the same compliance in the horizontal direction as the vertical. Structures that have relatively high stiffness in certain degrees of freedom can use offload systems that constrain that degree of freedom without adversely affecting the deployment kinematics. NASA Goddard Space Flight Center (GSFC) developed a *g*-negation support system for use on the solar arrays of the Global Precipitation Measurement (GPM) satellite [38] using air bearings and a fabricated flat surface. The two characteristics used to assess the performance of the solar array deployment system are the total deployment time and deployment path. The deployment time and path varied with characteristics of the *g*-negation installation specifically in the areas of hose routing and the

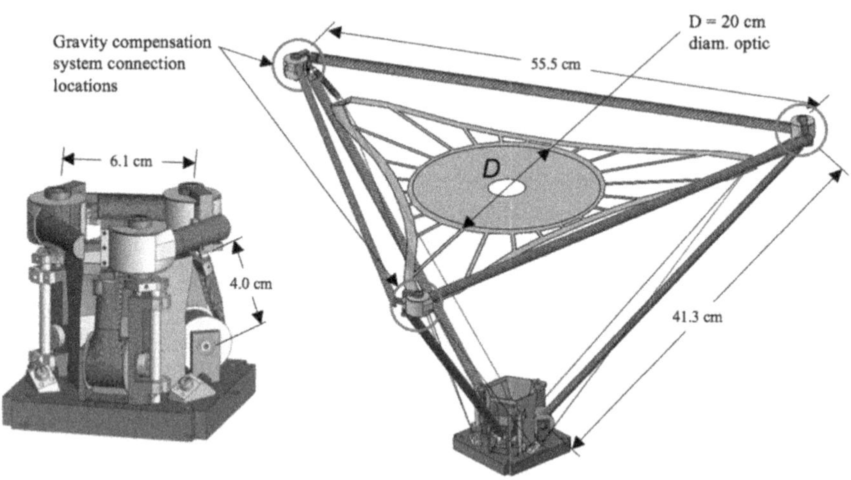

Fig. 4.18 The CDT structure. (Credit: AIAA.)

hose support alignment to the tables. Initial testing of the solar array design showed that the deployment time varied greatly with changes in the *pitch angle*, defined as the angle between the deployed boom axis and the deployment tables. Shoulder hinge angles outside −0.10 deg to +0.10 deg produced deployments that were longer than expected. Likewise, small changes in the roll angle of the shoulder hinge provided variable deployment paths.

The air hose configuration of the *g*-negation system, shown in Fig. 4.19, was nebulous in design and was photographed repeatedly from different vantage points to achieve a repeatable system. Slight changes in the air routing had the same effects as slight changes in the pitch and roll angle with respect to changing the deployment path and deployment time by generating friction at points in the system different from the desired baseline. By keeping accurate records of the pitch and roll angles, as well as the *g*-negation hose routing, the solar array deployment times and deployment paths were consistently within acceptable deviations. Telemetry graphs from the +Y Solar Array Wing observatory level pre-vibe and pre-ship deployments showed a three-second difference in deployment time, a 2.91 percent difference, and similar angular velocity characteristics between these two different deployment setups performed three months apart.

One advantage of the air pad system was that it accommodated first mode frequency verification testing. The qualification unit −Y Solar Array Wing was configured in the deployed position, and a direct measurement of the frequency was able to be performed by placing and releasing the side load and examining the

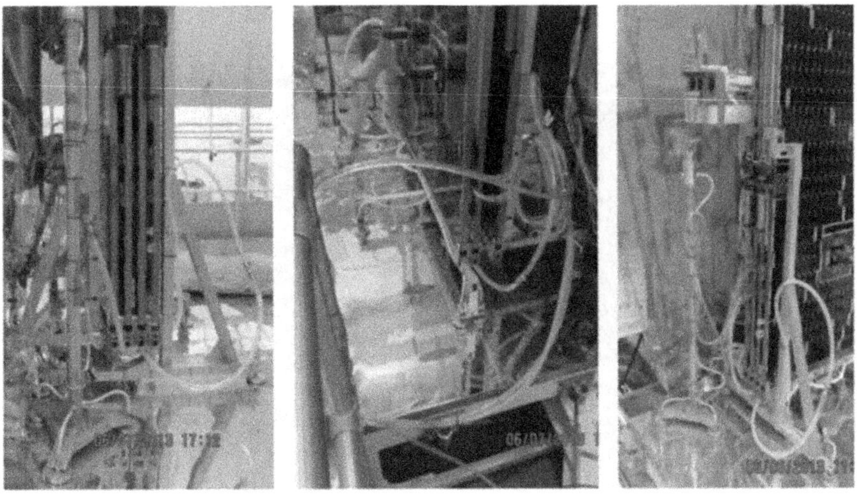

Fig. 4.19 Air bearing hose routing had measurable impact on deployment times for the GPM solar array [38]. (Credit: NASA GSFC.)

oscillating motion of the wing while supported by the air pads. Since the air pads induce almost zero friction and the measured value was found to be close to the predicted value, the analytical model was verified showing that the solar array meets its frequency requirement. This test method was found to be more straightforward than others that required suspending the solar array from cables and simply observe the free vibration response.

The AFRL Multi-Arm Radial Composite (MARCO) deployable reflector [39] had to select a gravity compensation system to be compatible with the two-dimensional deployment motions at a relatively high rate. The MARCO reflector consists of a plurality of independent folding radial arms repeated revolved about a central axis. The full-scale engineering unit test article consisted of 3 of 12 arms of a full reflector. A single arm describing the locations of the offload points is shown in Fig. 4.20. If one is testing a single arm, the orientation should be such that there is no motion in line with the gravitational vector. Even in this configuration, the displacements of the attachment point (range of approximately 1.5 m in two dimensions) required the offload cable attachment points to be traversable in both degrees of freedom. That is, a fixed offload cable imparted lateral parasitic loads that dominated the deployment. The motion profiles for the three-arm system include large vertical and horizontal motions, with no way to orient the structure to eliminate one component. The reflector subsystem was configured such that radial motion was perpendicular to the gravity field.

The limited separation distance from the structure to the laboratory ceiling where the offload cable would attach, coupled with the radial displacement, required the offload cable to be mounted to a low-friction cart system that

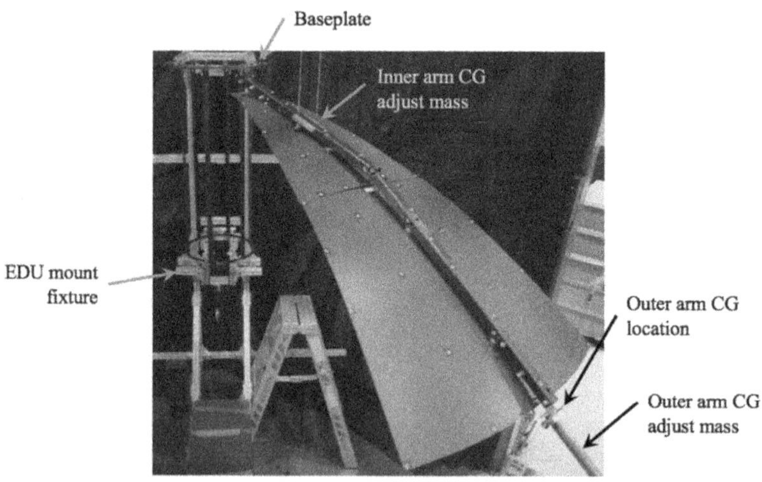

Fig. 4.20 The MARCO reflector arm required additional masses to relocate the rigid-body subassembly center of gravity (CG) points to physically accessible locations. (Credit: AIAA.)

freely traversed a linear rail. A pulley on the offload cart was used to mount a counterweight to provide the constant offload force. Stainless steel wire rope was originally selected as the cable material to give high margins of safety and the ability to quickly acquire and install commercially available end fittings rated for lifting applications. Subsequent calculations showed that the parasitic load created from mass of the cable moving from one side of the offload pulley to the other was on the same order of magnitude as the available deployment torque. The steel wire rope cables were replaced with a synthetic Spectra braid that rendered the cable length effects negligible while maintaining safe margins with respect to the working load.

The relatively high rate of the free deploying structure resulted in additional parasitic loading from the passive offload carts. The deployable arm has to pull an offload cart along with it. The cart inertial properties with the observed deployment velocities were used in the calculation of the parasitic loads induced, whose peak values were also on the same magnitude as the available deployment torque. The next development unit currently being designed and fabricate will incorporate rate control mechanisms for a few reasons, one being to mitigate this effect.

Another factor to consider when designing the gravity compensation system is access to the CG point. Some structures, like MARCO, can be discretized to individual subsystems that can be considered as rigid bodies. Offload connection points should be at these locations. Chapter 6 discusses the number and locations of connection points for a flexible (distributed) structure. Offset of the attachment point from the CG locations results in parasitic torques. The CG locations for the MARCO arms were not readily accessible due to the structure's gravity orientation and the overlapping architecture of the shell segments to compactly stow. The approach taken to minimize the parasitic loads was to add mass to relocate the CG location to a point that was accessible. The addition of mass does impact the deployment rate but was not overly significant in this case. The added masses are easily modeled and do not add significant uncertainty.

4.4.4 END OF DEPLOYMENT SHOCK LOAD TESTING

High-frequency shock inducing events typically occur at the deployment initiation and when motion ceases. Both events are related by the transition of energy from static to dynamic (or vice versa) in a relatively short time period. These shock events can be reduced by increasing the time through controlled deployments. The presence and concern about damaging levels of end of deployment shock are obvious for the class of "free-deployable" structures where the rate control mechanisms are intentionally omitted to satisfy extreme volumetric and/or mass constraints. Implementation of HSC components can impart shock loading even with overall deployment rate control due to the snap-through [40]. Figure 4.21 shows the tape spring reaction moment versus deployment angle. The sharp change in reaction moment magnitude at the end of deployment

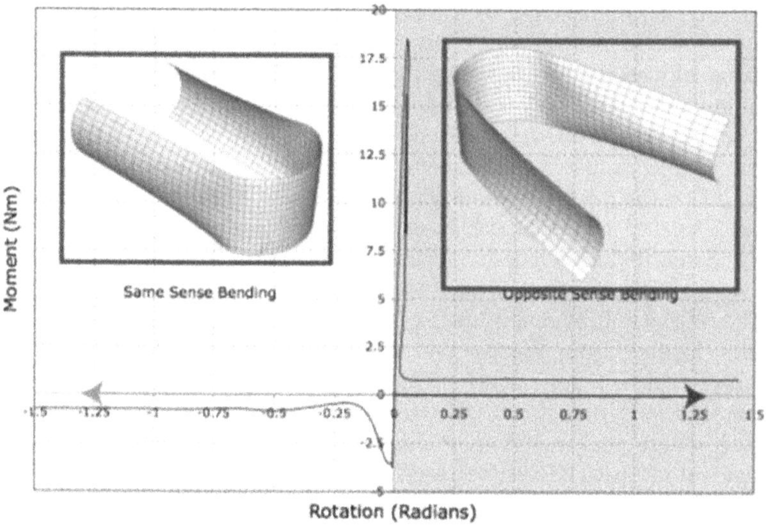

Fig. 4.21 HSC components such as composite tape spring hinges experience a rapid, shock-inducing event when transitioning to the deployed state, even with quasi-static controlled deployments [40]. (Credit: AIAA.)

indicates the shock event due to the sudden release of strain energy. The impact of these events on the host spacecraft control system and structural integrity of the deployable as well as surrounding structures must be assessed.

In a broad sense, this impact needs to be characterized with some standardized methodology, such that system integrators will have high confidence that the deployment of the structure will not damage surrounding structures. This shock must be mitigated by the inherent design of the structure and its deployment, but it must also be characterized to bound the shock and energy imparted onto the spacecraft.

The primary issue is how to test and measure the shock loading imparted onto the spacecraft from snap-through, rebound, and other shock events during deployment. It is possible to first bound the amount of energy released *a priori*, and from this, a rough order of magnitude in energy during a shock event may be gleaned. At a minimum, the snap-through shock event can be bound by the strain energy released over the time it occurs. The energy is finite and easily modeled. The time is measurable via experiment or can be computed via simulation.

With a free deployment, structure inertial and deployment speed are key factors with shock loading. When initial deployment reaches the end of travel, the inertial mass rapidly reverses during a rebound event. The event is bounded by the kinetic energy of the inertial masses, while the rebound time is primarily governed by the structural stiffness between the deployed mass and the host spacecraft.

Shock is complex and highly dependent on boundary conditions and structural interactions. Traditionally, shock testing is performed and measured on the actual spacecraft or a structural facsimile. However, in the research and development environment for a deployable structure, the problem is that there is not a host spacecraft to test on. When one is considering a HSC structure or free-deployed structure, shock still needs to be tested and measured in some meaningful way to quickly and simply convey the general magnitudes of energy generated from the deployment of the structure.

Shock magnitude is directly dependent on time. With a stiff interface, strain energy travels quickly through it, which increases the peak shock magnitude. With a compliant interface, strain energy travels more slowly and therefore reduces the peak shock magnitude. This concept is used on CSA Engineering's SoftRide [41] to reduce vibration magnitudes during launch. This same concept may be applied to address shock events from the deployment of a structure with small passive mechanisms.

One approach is to measure the shock response on a "surrogate host" platform with some measurable compliance closely resembling the envisioned host's local mounting conditions and structure. Then, measure the shock at the structure-host interface and compute an industry-standard shock response spectrum (SRS) curve [38]. Then use the SRS to convey a quick and simple worst-case scenario.

However, the SRS methodology has some limitations. It is an empirical tool that attempts to capture the complexities of shock loads. The basic model involves breaking up the response into frequency bins. Each bin contains a spring-mass-damper (SMD) system tuned to that bin, where the tuning has estimated parameters based mostly on experience. An SRS basically computes the maximum amplitude of each SMD when excited by the test shock time history. Significant amounts of information are lost from this process. When one is trying to reverse an SRS to test a structure, infinite time histories can satisfy it. So, a test like this is not definitive in any way; rather, such a test is a gross approximation. This is why it is best to test the actual test article on the actual host structure. However, there still has not been a significantly better way to model and convey shock response than with the SRS methodology.

FEM can perform this function, but it is computationally expensive and requires a model of the host spacecraft and deployable structure. While reduced-order models can be used to lighten the computational load and provide a clean separation of engineering tasks, this further degrades the accuracy of the computed shock response and often still requires a physical test on the actual host spacecraft and structure.

SRS is commonly used to describe the launch environment and separation events, but no literature has been found on research or use-cases for dynamically deployed structures or quantifying high-energy snap-through events on a host spacecraft. This seems to be primarily due to most deployable structures being quasi-statically deployed, as mentioned earlier. As a result, the use of SRS to characterize shock with deployable structures seems to be novel and unexplored.

Based on the history of SRS, it seems as if SRS is the logical choice to convey this type of information quickly and easily.

Shock response is dependent on both stiffness and mass. The idea of testing with a surrogate host serves to model the local interface stiffness, where the shock energy travels directly through, but does not fully capture the influence of, attached mass, structural stiffness, and non-structural components. If the mass of the host structure is much larger (three orders of magnitude) than that of the test article, the mass influence on shock response is minimal and may be modeled as a rigid interface. However, if the mass of the host structure is not significantly larger than the deployed mass, a host mass simulator should be attached to the surrogate host and needs to have inertial and compliance characteristics that are close to those of the intended host spacecraft.

A typical shock test involves attaching the test article to the host spacecraft. Shock sensors are placed close to the interface boundary condition (within a few centimeters), where the load transfers between the test article and host structure. Any further increase in this distance would cause the modes of the host structure to become more influential, and the measurements could be misleading.

4.4.5 METROLOGY SYSTEMS

This section presents the various available metrology technologies that are useful for deployment testing of large ultra-lightweight structures. It is assumed that sensor sensitivity selection and placements were performed during the test design process, and this topic is not covered here. A brief overview of each metrology system is presented along with the main considerations and impacts on the deployment test.

4.4.5.1 TRANSDUCERS

A *transducer* is defined as a device that converts variations in a physical quantity, such as force or acceleration, into an electrical signal. Data acquisition systems measure the electrical signal and transform the value by the transducer calibration factor (also known as *sensitivity value*) to present the measurement in its physical units. Transducers (or *sensors*) that are commonly used in structure deployment testing are presented below. All devices are available in versions compatible with harsh environments, including vacuum and temperature ranges.

Load Cells

Load cells are categorized as being designed for static forces or dynamic loads. There is some flexibility in the sensor specified frequency bandwidth for each type with minor reductions in accuracy. This could be useful to potentially not require both types in the load line. Data sheets for static load cells typically do not list a bandwidth for which the specified accuracy is valid. But static load cells can be used in dynamic applications up to around 2 Hz. Similarly, dynamic load cells will have a bandwidth specification that calls out the lower frequency bound. If the sensor has a sufficiently large discharge time constant (some are >50 sec), loads can be accurately measured well below 1 Hz. It is recommended that the test engineer verify the sensor accuracies with the proper calibrated transducer prior to using it outside the specifications.

Multiple degree-of-freedom load cells are commercially available, but the majority are designed for axial force only. Off axis loads increase measurement error and can damage the load cell. Implement connecting fixtures such as spherical joints or flexures, and verify that the load cell is subjected to only axial loads for the full range of motion of the deployable structure.

If using a multiple degree-of-freedom load cell where the multiple axes are mechanically coupled, note that since noise is present in all the individual sensor elements, the axial load is amplified due to the matrix multiplication. For example, if all the signals have noise with similar magnitude, the noise of each channel is present in the calculation of the resultant load. It more or less is amplified by a factor of six.

A *torque cell* is a transducer that measures the static and dynamic torque of a connected rotating shaft. These devices are useful for characterizing the running torque required to deploy the structure and for determining the relationship between motor power draw and applied torque.

Accelerometers

Typical accelerometers sense only changes in acceleration and are therefore insensitive to the gravity field. For a nearly constant rate deployment, the accelerometers would detect only the vibrations during the deployment. A class of available accelerometers has a bandwidth down to 0 Hz (often called *DC responding*) or steady state, and these accelerometers are sensitive to the gravity field. These sensors can be used to measure tilt angles of the deployable structure. They are also suitable for dynamic measurement, with common upper frequency bounds on the order of 1 kHz.

Drive Motor

If the structure utilizes an electromagnetic motor to drive the deployment, the motor can be utilized as a transducer. There is a direct relationship between the motor current and torque output. This can eliminate the need to add a torque cell in series to verify torque margins are satisfied during test.

As long as the motor is carefully sized to the application in respect to torque and speed, a stepper motor can be commanded to move to and hold at a desired position without any feedback sensor (an open-loop controller). Recording and referencing the commanded steps, with knowledge of the motor output gearing ratio, can be used to give the structure position during the test. This measurement is not absolute, and position knowledge can be lost in the event of power loss or errors within test software. The structure could be commanded to a known position, and the position tracker is then reset.

Rotary Encoders

Rotary encoders can be instrumented at hinge-lines during a test to record the deployment angle versus time. There are two main types: incremental (relative) and absolute. The output of *incremental* encoders provides information about the *motion* of the shaft, which is typically further processed elsewhere into information such as speed, distance, and position. An incremental encoder accurately records changes in position but does not power up with a fixed relation between encoder state and physical position. Devices controlled by incremental encoders may have to "go home" to a fixed reference point to initialize the position measurement.

An *absolute* encoder maintains position information when power is removed from the system. The position of the encoder is available immediately on applying power. The relationship between the encoder value and the physical position of the controlled machinery is set at assembly; the system does not need to return to a calibration point to maintain position accuracy.

Linear Displacement Sensors

There are various different types of linear displacement sensors. Laser displacement sensors are advantageous due to their relatively high accuracy and are noncontact. The main downside is the limited measurement range, which is typically less than 1 m. Displacement sensors that require contact with the structure include cable position transducer (CPT), also called *string potentiometers*, and linear variable differential transformers (LVDTs).

CPTs consist of a string mounted to a spring retractable reel coupled with a rotary encoder. The force applied by the CPT on the structure under test can significantly affect the deployment rate of the lightweight structure. These devices have measurement ranges up to tens of meters and can support high cable feed rates up to 100 *g*s. Examples where CPT frequency response can be an issue are applications that require cable accelerations in excess of 100 *g*s and environments using analog-output CPTs with very high-frequency, low-displacement motion that tends to cause dithering wear of the potentiometric element. During extreme high accelerations, the cable reel's inertia and the power spring's insufficient torque could cause the reel to rotate to a point where the displacement cable goes slack.

LVDTs can perform high-accuracy displacement measurements at high rates (1 – 10 kHz), but they are limited to displacement ranges on the order of 1 m. These devices consist of a shaft within a cylinder that converts the mechanical position into a proportional electrical signal. The center shaft can be mechanically attached to the structure or held against the structure by an axial spring. If the LVDT is spring loaded, the force applied must be accounted for similarly to the CPT. Applications where the LVDT is fastened to the structure will have the added inertia effects, and the attachment must be designed such that non-axial loads are placed on the transducer.

4.4.5.2 VIDEOGRAMMETRY

Videogrammetry is a well-established metrology methodology [42, 43] that expands photogrammetric techniques into the time domain, yielding three-dimensional point-cloud data of the article under test. Systems use multiple video cameras connected to a centralized controller. These cameras are positioned around the test structure, which has a number of targets mounted at desired locations. Basic triangulation calculations are made using the two-dimensional images from all the cameras to give the spatial location of each target centroid with respect to an established coordinate system. The triangulation computations are made in real time; no post-processing is necessary to obtain the target centroid data.

Targets are nominally spherical to produce a circular two-dimensional image at any camera incidence angle. Flat circular targets may be used (but are not advised for situations where the incidence angle is high), and such targets yield elliptical images that increase centroid location errors within the frame along the semi-minor axis. This often results in failed target tracking when one is triangulating the target's position within the system error bounds. Depending on the system, targets can be passive or active. *Passive* targets are simply foam or plastic spheres covered in a reflective material. *Active* targets are also spherical but are illuminated by an internal LED. The LEDs are illuminated at precise time intervals to provide unique target identifications without the need for model-based identification. This is beneficial when one is capturing irregular movements and when markers move independently from each other. When passive targets are used and can move independently from each other, the system can calculate the spatial centroid location of all targets for each frame capture, but the system does not maintain unique identifications. That individual tracking must be performed in post-processing.

Videogrammetry systems that do not use discrete targets are also available. Such systems are advertised as "markerless," and they identify key features of the test article being observed. To minimize the position error, care must be taken to ensure that the visual features being tracked remain fixed with respect to the number of pixels observed by the cameras. Therefore, the features should not physically deform in a manner such that the calculation of the centroid, or other tracking metric, does not yield the true position during the test. For the same reasons, the feature must also reflect light to the cameras consistently throughout the deployment motion.

Various commercially available products that provide the complete system that consists of cameras, controller, and software are available (do an Internet search of "motion capture systems"). Commercial providers were initially focused on the entertainment industry to support human motion capture, and the software was geared toward that specific application. Due to the increased adoption by the engineering and science community, software toolkits that are compatible with common engineering applications such as Mathworks MATLAB and National Instruments LabView are now available.

Videogrammetry is highly attractive for deployment testing of LULS for a number of reasons:

- *Non-contact metrology for passive targets.* Passive videogrammetry targets can be readily fixed to locations on the structure with no cabling necessary and with the addition of relatively minimal mass. As mentioned in "Transducers" section, earlier in this chapter, cabling can produce measurable mass and stiffness effects on the deployment response. Instrumenting the test article with wired transducers adds setup time and requires the attachment of cable management features that must be removed later. The stiffness and effects also must be accounted for in the model, which both increases the complexity and computation time and adds other uncertainty parameters to the overall system.
- *Simultaneous measurement of large number of targets.* Common videogrammetry systems can sample hundreds of targets. This allows for increased comparisons to the model, which reduces overall uncertainty. The number of targets and the data acquisition rate are limited by the system's data throughput capabilities. Camera resolution and acquisition rate are programmable features.
- *High accuracy and precision over large volumes.* The achievable absolute accuracy is a function of the arc angle, the number of pixels within each image, the number of cameras that view the target for a given frame, and the accuracy of the calibration object. Volumes up to 10 m^3 are readily obtainable with submillimeter precision. The target sphere diameters should be maximized subject to the packaging constraints of the structure in the starting stowed configuration. The number of cameras is related to the size of the observable area while maintaining accuracy. At some point, too many cameras can make the calibration unstable, whereas too few can make the measurement inaccurate.
- *Targets can be hidden or undetectable for period of time.* As mentioned above, the target tracker software provided by the system developer does not retain knowledge of an individual target from one frame to the next. For each frame, the spatial coordinates are calculated for each detectable target and recorded in no particular order. To assign identification and track each target throughout the test, post-processing is required. But, this is beneficial for applications where targets can go into/out of view during the deployment process. LULS are typically stowed into compact configurations and/or restrained within a canister where the majority of targets cannot be viewed.
- *Lower acquisition bound is 0 Hz (static).* The acquisition bandwidth for videogrammetry goes from 0 Hz (static or DC measurements) to the upper bounds described above. This metrology system can be used for both very slow and

rapidly deployable structures. With this feature, the same camera setup configured for deployment testing can be utilized for pre- and post-deployment static measurements. This can eliminate the need for a photogrammetry system for static shape measurements and a suite of transducers and data acquisition system for deployment testing.

The Air Force Research Laboratory, Space Vehicles Division, implements the Vicon MX videogrammetry system. Rather than using the Vicon software, which tracks rigid objects of three or more targets, custom MATLAB software was created to track individual targets from the raw target data. The procedure used in their deployment tests is presented. For each deployment test

1. The videogrammetry system is powered on and allowed to naturally come to normal operating temperature. At least one hour is given to ensure the temperature has stabilized. Measurable variations are present due to thermal deformations of camera components.
2. Ideally, calibration is performed by moving the calibration wand through the volume that the deployable structure occupies in both the stowed and the *deployed* state. If the test article cannot be removed from the test volume, the structure is stowed, covered, and calibrated as close as possible to the test article. The calibration defines the coordinate system for the data capture. If necessary, an assessment of calibration uncertainty may be observed by performing multiple calibrations and measuring changes in distances between fixed targets throughout the test volume.
3. A data capture is performed on the structure in the stowed state. This gives the initial locations of the targets. Any targets that cannot be observed in the stowed state can be added manually using CAD geometry models.
4. The structure is deployed while acquiring target data with acquisition buffers of at least 1 sec before and after deployment.
5. The initial shape static data and deployment dynamic data are exported in individual text files.
6. The two text files are imported into MATLAB, where a custom script executes the following post-processing procedure:
    a. The stowed structure data is imported and averaged to determine the initial location of each target, and a unique identification name is assigned. Any targets that cannot be observed in the stowed state can be added manually using CAD geometry models.
    b. The deployment data is loaded. Recall that the target data is not sorted and that the data vary from frame to frame. For each target, the distance to all targets in the next frame is computed. The minimum distance is assumed to be the match of the current target. If this minimum distance is outside a predefined bound (0.5 mm, for example), it is assumed that the target was not observed in that frame capture and the location for that target is

assigned a NaN value. If it is within the bound, the target location is updated and becomes the baseline for the distance search for the next frame. The bound is determined based on the anticipated deployment speed and acquisition rate.

c. False target identifications from reflections or other unknown sources can be present in the data. These targets are retained and tracked for review by the engineer.
d. After all frames have been processed, any NaN values are replaced by linear interpolation.
e. The transformation matrix from the current coordinate system to the reference coordinate system (typically, the coordinate system used in the finite element model) is calculated. The time coordinate data are transformed to the reference coordinate system.

Within Airbus Defence and Space Netherlands (ADSN), a videogrammetry system to measure the motion of spacecraft subsystems during its deployment has been developed. Commercially available tracking systems were considered, but it was decided to develop a custom system mainly to drastically increase the measurement volume. The specified volume of commercial systems ranges on the order of magnitude of 10 m^3. ADSN required a system that could perform measurements over the test volume at least 2500 m^3 such that two larger deployment rigs in the ADSN cleanroom could be observed with the same metrology setup. One of the advantages of this increase in measurement volume is that deployment tests are always performed within the measurement volume of the system. No readjustments of the system are needed, avoiding a time-intensive and complex recalibration of the system. This increased test volume is, however, at the cost of accuracy of the measurements; where commercially available systems can perform measurements at sub-millimeter precision, the developed system is capable of measuring markers with a precision of 5 mm or better. Although the accuracy is an order of magnitude lower, it is adequate for the purpose of measuring typical ADSN deployment systems. Note that the cameras can still be zoomed in to focus on the test article, increasing the accuracy of the measurement. This, however, requires a recalibration of the system, as the configuration of the system has been modified.

The developed system is based on the stereovision object-tracking principle, extracting the 3D position of markers from digital images. To do these measurements, four cameras, combined with a large number of markers attached to the test article, are used. The four high-resolution cameras—fitted with a Near Infrared (NIR) bandpass filter—are located in the corners of the cleanroom ceiling. The cameras work on a global shutter principle and are trigger controlled, ensuring each camera captures an undistorted image at exactly the same moment. The used lenses have been selected for their high field of view (FoV) of 108 deg horizontal and 79 deg vertical. This setup allows each camera to capture the required measurement volume. The used markers emit NIR light using two 850 nm LEDs encapsulated with a diffuser of 40 mm. This marker size is just small enough to be fitted on the test article for the stowed configuration but is still visible by the cameras when located at the far end of the cleanroom at a distance of 35 m. The diffuser ensures a constant circular shape and light intensity, independent of the incidence angle of the camera. Markers are mounted either to the zero-*g* suspension of the solar array or directly to the test article. These active markers require a dedicated power supply, which can be either a battery pack or an external power source, dependent on the deployment system's complexity and sensitivity.

Calibration of the system is required on both the lens level and the system level. For the calibration of the lenses, the 2D-calibration principle is employed: For each camera, many images of a checkerboard pattern under different angles are captured. The calibration software uses these images to determine the intrinsic camera parameters, among which the most important are the lens properties and the lens distortion. The system calibration is based on the 0D calibration principle, which requires a single marker moving through the entire measurement volume. Based on these measurements, the calibration software is able to determine the extrinsic camera parameters, which are the position and orientation of all cameras with respect to each other. A recalibration of the intrinsic parameters is required only if the lens configuration is altered, and a recalibration of the extrinsic parameters is required only if the camera configuration is altered.

Apart from the in-house development of the hardware, the tracking software has also been developed within ADSN. The software uses the commercially available software package MATLAB and its image processing toolbox. The toolbox includes the calibration, processing of the video-recordings, tracking of the 2D trajectories of the markers, the triangulation algorithms, and the visualization of the measurements. All the data processing is done after the measurements. This in contrast to commercially available systems that apply real-time processing of the data during the measurement. This approach reduces complexity of both hardware and software but results in a tremendous amount of data to be stored, which is in the order of magnitude of 30 GB of camera recordings per minute. The data processing starts with the compression of recorded videos. In this step, contrast thresholding is included, to omit all background light from the videos. The processed videos are further subjected to the 2D tracking software. Once each marker has been assigned manually, autonomous tracking is possible with almost no human interaction. In order to achieve autonomous tracking, motion prediction algorithms are used aided with Kalman filtering. Marker motion prediction is performed for each frame based on the data of the previous frames, i.e. the 2D-pixel-coordinate as well as the 3D-position. Besides motion prediction at marker level, motion prediction on body level is performed, which eases the computation required for tracking of the markers. The combined motion prediction and inverse tracking results in a robust algorithm capable of handling with blind spots, overlapping markers and other anomalies.

Robustness of the tracking algorithm has been improved by the implementation of dedicated algorithms that solve the problems of marker occlusions, overlapping markers, and other anomalies. The significant number of marker occlusions is caused by the deployment rigs, whereas the issue of overlapping markers mainly occurs when structures such as solar arrays are in the stowed configuration. To allow stable and robust tracking, these algorithms are indispensable. The most effective algorithms are based on inverse tracking: If the 3D position can be derived by at least two cameras, the 3D data of the marker can be used to determine the 2D pixel coordinates for the cameras for which no marker could be found.

After the deployment is tracked, the data are processed, and the user can study the deployment data in several ways, either from graphs or a 3D visualization of the deployment. A 3D visualization of a typical three-panel solar array deployment is shown in Fig. 4.22. Typical graphical results include the marker trajectories and the position, angular orientation, and angular velocities of the panels.

4.4.5.3 VIDEO MOTION ANALYSIS

Video motion analysis is similar to videogrammetry. The central difference is that with video motion analysis, tracking of identified targets is performed on images collected from a single video camera. Within the FoV is a reference scale that is used to transform the motions observed in the video to the actual physical displacements. Sophisticated commercial software packages to perform the motion analysis are available. No additional software (other than video-viewing software) is required for simple tracking analysis that can be performed by comparing the points of interest to the reference scale. A parallax correction factor must be applied to the measured distances to account for the perceived travel versus actual travel. *Parallax* is a displacement or difference in the apparent position of an object viewed along two different lines of sight, and it is measured by the angle or semi-angle of inclination between those two lines. This is due to the offset of the scale from the centerline of travel. The camera should be positioned such that the viewpoint is perpendicular to the structure deployment travel vector.

a) b)

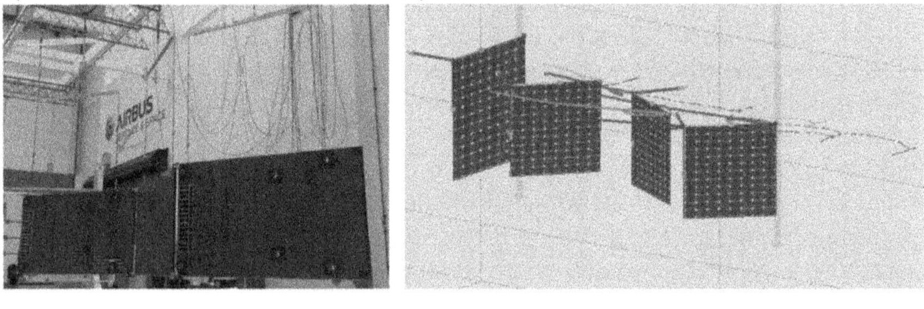

Fig. 4.22 A three-panel solar array during deployment testing (a) and the 3D visualization reconstructed from the deployment measurement (b). The arrow indicates a tracking marker on the test article. (Credit: ADSN.)

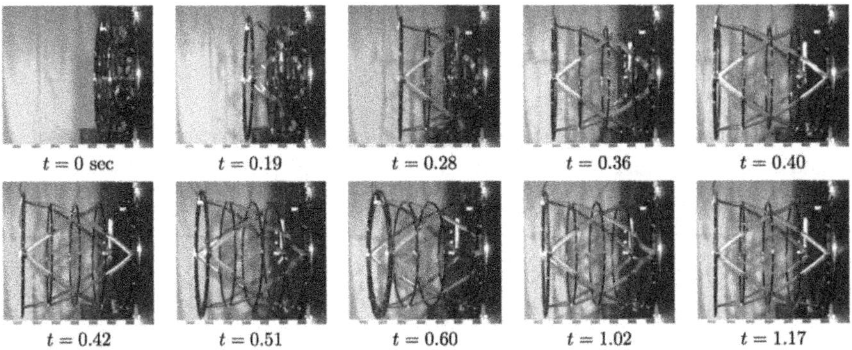

Fig. 4.23 High-speed camera still shots of a deployable sparse isogrid structure [44]. (Credit: AIAA.)

The time component is obtained by knowledge of the camera frame capture rate. For fast deploying structures, such as the sparse isogrid [44] shown in Fig. 4.23, a high-speed and high-resolution camera can be used. Any video camera can be used for this metrology.

4.4.5.4 LASER TRACKERS

Laser trackers consist of a stationary head unit that emits the beam laser and measures the position of a target that is mounted to the structure. These systems are designed primarily to survey multiple targets on a static structure, or the test technician moves the target to multiple locations while the laser head tracks the target. This tracking capability can be utilized to measure the position of a single target during the deployment process. The target velocity must be within the capabilities of the tracker unit. The laser head needs to have an uninterrupted view of the target so that it does not lose the signal. Because of this, a laser tracker is limited to tracking one target only.

4.4.5.5 CABLE MANAGEMENT

Independent of the transducer type, the sensor cabling can have a profound effect on the deployed dynamics of lightweight structures [45, 46] due to added stiffness, damping, and mass. The same effects are present on the structure during the deployment as well. The research performed on cable effects showed that accurately predicting the influence is extremely difficult. When one is instrumenting the system for deployment testing, the number of sensors used should be carefully considered due to the associated issues that could be created with the added cabling. One purpose of the test is to verify the analytical models. More data from the test naturally drive down the uncertainty, which directs the engineer to maximize the number of sensors. But, the compounding mechanical effects on the deployment responses adds uncertainty, which can exceed the benefit from the added information obtained.

The overarching goal is to minimize the parasitic forces applied to the structure from the cabling during the deployment. One method is to route the cable along the structure to a fixed (non-moving) section. When one is traversing joints or extensional members, ensure there is sufficient slack in the cable to accommodate the range of travel. Another method is to have the cable immediately leave the structure from the sensor location. This eliminates the issues described above, but now the cable has to deploy along with the structure. Having the structure pull the cabling along (dragging along the floor) should be avoided. Suspending the cables from the gravity compensation system, or utilizing an independent system that operates in a similar manner, is a recommended solution. With either method, care must be taken to ensure that it is not possible for the cable to snag.

Cost and schedule permitting, a deployment test with a single sensor could be performed prior to fully instrumenting the structure. This gives a baseline for the deployment to compare with subsequent tests and to give quantifiable evidence of the sensor cable effects.

4.5 CASE STUDIES

The following case studies present in further detail specifics of the deployment test program for two classes of LULS. The first case study discusses the flight qualification of large, rigid panel solar arrays designed to deploy at a relatively slow rate. The second case study examines testing of a prototype deployable column implementing HSC members with no rate control.

4.5.1 RIGID PANEL SOLAR ARRAY

A solar array wing assembly (SAWA) in this example has five panels, each approximately 144 in. × 72 in. wide, attached to the spacecraft via a yoke and boom. The wing is held in place via release devices in eight places. Once the panel constraints are released, the wing deploys via spring torque at each of the seven hinge-lines, which deploy independently (they are dynamically but not kinematically linked). The deployment rate is limited by dampers, one per hinge-line. To control the motion, a delay hook prevents deployment of the short boom until the yoke has deployed approximately 15 deg. The wing is decelerated and "locked" in the deployed position in a gradual process as the damped hinges develop increasing mechanical advantage over their deployment. Figure 4.24 shows rigid panel solar arrays mounted to various spacecraft.

The dynamic motion during deployment is primarily sensitive to the hinge spring torque, the varying mechanical advantage of the hinge cam action, the damper deadband and damping characteristics, and the range of delay angles over which the delay bracket releases the short boom. Other parameters such as the harness bending torque, the kickoff springs, and differences in panel weights and mass centers are included in the deployment model but have only a minor effect on deployment motion or lockup loads.

The dynamic model is created and solved in Automatic Dynamic Analysis of Mechanical Systems (ADAMS), a standard software used for dynamic analysis of large-motion, multi-body systems. Because of the numerous trajectories and associated loads possible for the wing, a Monte Carlo analysis is performed to determine the worst-case behavior.

A *Monte Carlo analysis* is a statistical summary of numerous deployment simulations with parameter values randomized. This analysis gives the best estimate of both nominal and worst-case performance for multi-panel deployments which have many coupled and highly non-linear degrees of freedom. A Monte Carlo analysis is also effective for representing the full range of deployment trajectories. This allows stay-out zones to be defined for wing deployment.

Fifteen hundred simulations were performed with over 150 parameters varied over their worst-case range. The values are randomized uniformly over the full range of predicted behavior (not restricted to "worst case" values) except for certain variables where the distribution is expected to be normal (Gaussian) and where a uniform distribution would be too conservative.

The test program overall focused on doing more extensive testing at the component level, and then workmanship testing at the full array level was done. The testing described herein is intended to verify the workmanship of the hinge, panel, and wing assembly.

The primary deployment mechanisms for the array are the hinges, and thus the hinges are tested most extensively. Resistive torques from the wire harnesses are measured in qualification testing, and then torque margins are confirmed in each acceptance test by measuring torque versus angle. Each flight hinge is tested with the flow shown in Fig. 4.25, which confirms acceptable spring torque, friction, and performance following vibration and thermal vacuum testing.

Each panel-level test article is composed of solar cells bonded to substrate with all wiring completed. Substrates are tested prior to cell bonding for strength. Manufacturing verification tests should demonstrate acceptable performance at the panel level over the specified range of mission requirements. The tests should

Fig. 4.24 Lockheed Martin rigid panel deployable solar arrays. (Credit: Lockheed Martin.)

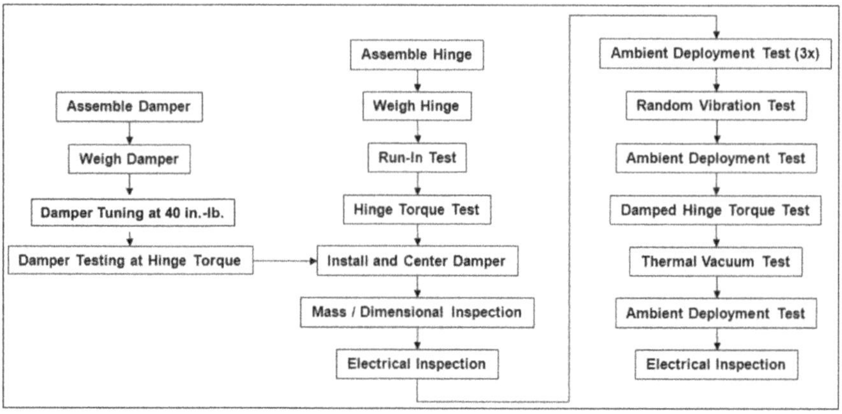

Fig. 4.25 Assembly and test flow diagram for hinge qualification. (Credit: Lockheed Martin.)

reveal inadequacies in the manufacturing process such as workmanship or material. The performance measurements used for a baseline should have enough fidelity to identify any degradations or parameter drift even if the data are within specified limits. Test levels specified are per the predicted flight level plus acceptance margins.

A wing-level test article is composed of five panel assemblies integrated together with yoke and boom assemblies via a pair of hinges at each hinge-line for Interpanel Hinges (IPH) and Boom to Panel Hinges (BPH). The yoke to boom hinge-line consists of one 180-degree hinge, and the boom to spacecraft hinge-line consists of one 90-degree hinge. Acceptance tests are intended to demonstrate the flightworthiness of each deliverable item. Acceptance tests should demonstrate acceptable performance at the wing level over the specified range of mission requirements. Manufacturing verification and acceptance tests for the panel and wing assemblies are listed in Table 4.1. The tests should reveal inadequacies (such as workmanship or material) in the manufacturing process. The performance measurements used for a baseline should have enough fidelity to identify any degradations or parameter drift even if the data are within specified limits. Acceptance test levels specified are per the predicted flight-level plus acceptance margins.

4.5.1.1 FULL DEPLOYMENT

The SAWA is completely deployed from a stowed to a fully deployed configuration, as shown in Fig. 4.26. Since the panels and hinges are tested at the component level, and the true flight deployment is proven by the Monte Carlo analysis, the deployment test at the solar array wing level is a demonstration that there are no unknown interactions or major issues with the deployment. The SAWA is attached to

TABLE 4.1 MANUFACTURING VERIFICATION AND ACCEPTANCE TEST PLAN FOR PANEL AND WING ASSEMBLY

Item	Test	Panel Level		Wing Level	
Environmental					
1	Thermal cycles at ambient air pressure	A			
2	Spacecraft environmental tests followed by release of hold downs and first-motion deployment			A	
Functional (before and after each environmental test)		Pre-Thermal	Post-Thermal	Pre-Acoustic	Post-Acoustic
3	Visual inspections for mechanical integrity	V	V	A	A
4	Illumination testing		V	A	A
5	Continuity and isolation testing	V	V	A	A
6	Deployment tests (full wing deployment to hinge locking)			A	A

V = Manufacturing Verification Test
A = Acceptance Test

spacecraft interface simulator and a calibrated low-friction deployment rail system to simulate a micro-gravity environment for deployment at ambient conditions. The deployment rail allows two degrees of freedom motions as the array deploys. The friction in the system in the deployment direction is compensated for by a slight tilt in the rail such that the gravity force equals the friction force. Each panel is offloaded with a spring-loaded system with a load cell to ensure the correct mass is offloaded. Obviously, with the size of the panels, air drag

Fig. 4.26 SAWA undergoing deployment testing. (Credit: Lockheed Martin.)

affects the deployment time and motions. For large rigid panel deployments, air resistance can be a significant factor and change the deployment dynamics. Gravity effects remain the dominant factor, however. Interactions with the offload system, and changes in the deployment response, have been observed because of differences in the properties due to gaps and frictional changes in joints.

At fully deployed configuration, the total offset of the SAWA when measured from the outboard panel relative to the 90-degree hinge/solar array drive plate centerline is measured and adjusted if necessary.

4.5.1.2 COMMAND RELEASE/FIRST-MOTION DEPLOYMENT

After spacecraft dynamics testing was completed, the SAWAs, again offloaded to simulate micro-gravity environments, are commanded to release their shear tie hold downs and allowed to deploy for the first 10 to 20 degrees of deployment to verify release will occur on orbit. The first motion deployment test is shown in Fig. 4.27.

4.5.2 SPARSE ISOGRID STRUCTURE

AFRL Space Vehicles Division developed the high-compaction, self-deploying sparse isogrid structure shown in Fig. 4.28 in both the a) deployed and b) stowed configurations. The general design aspects of the sparse isogrid column consist of a series of aluminum batten rings separated along the length by a

Fig. 4.27 First motion deployment of the solar array wing. (Credit: Lockheed Martin.)

crossing helical lattice pattern of HSC tape springs, indicative of an isogrid structure. The sparse isogrid column utilized HSC tapes to enable a deployment free of any sliding friction mechanisms and sources of dead play. One of the main objectives of this effort was to both determine the feasibility of and characterize the free, uncontrolled deployment of the structure to provide support that this approach could be implemented in a spacecraft program with acceptable risks. Using only internal strain energy, the deployment trajectory is linear, deterministic, and repeatable.

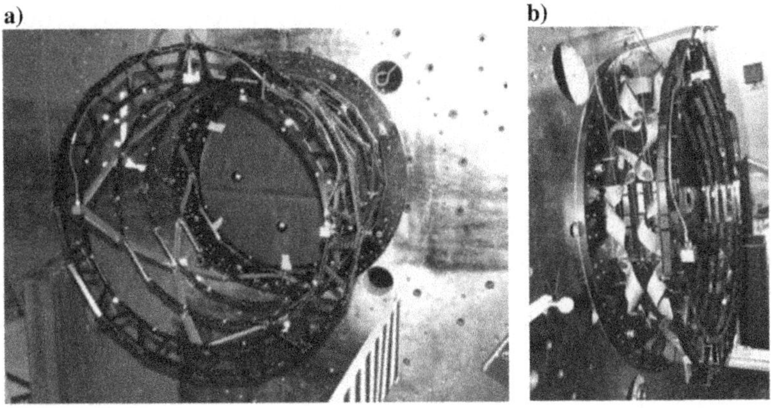

Fig. 4.28 Assembled test article mounted to backstop and gravity compensation system. (Credit: AFRL.)

The test article is mounted horizontally to a 1.8 m diameter, 10 cm thick, vertical steel plate. Two gravity compensation lines are attached to both sides of the horizontal mid-line of each ring, rather than the center of mass, since axial rotation of the structure during deployment is not expected to a significant degree. Each gravity compensation line is made of a lightweight braided Spectra filament, attached approximately 6.4 m above to a set of soft linear springs providing the offload force.

Only a tension restraining line on the tip-interface ring is needed to secure the stowed configuration, as shown in the lower part of Fig. 28b. The release of the test article is initiated by simply cutting the restraining line.

The experimental setup included a videogrammetry system, a high-speed camera, and a set of accelerometers. A Vicon T-Series videogrammetry system was set up with five T-160 visible-red cameras, arranged in a circular pattern about the axis of the column, viewing into the interior. Retro-reflective markers were placed primarily on the tip-interface ring and the base. The system was used to measure and compare the shape of the structure before and after multiple deployments, as well as to capture the dynamic motion of the deployments at a 250 Hz frame rate. Next, a Phantom Miro M110 high-speed camera was placed facing the side of the column to capture and to observe the minute details of free-deployment at a frame rate of 1600 fps. Finally, four PCB 311B1150G tri-axis 50 g accelerometers were placed on the structure, three on the tip and one at the base, to measure dynamic shock loads at a 2000 Hz sample rate during deployment and to determine structural vibration modes before and during the experiment for both model validation and checking for structural damage.

Before the first deployment, the test plan starts with initial static shape measurements with the Vicon system over multiple calibrations. Impact hammer tests were conducted before the first deployment and after each deployment to measure and detect changes in the structural vibration response, where changes would typically indicate damage. Three deployments were performed. Each deployment was captured concurrently by the Vicon system, high-speed camera, and accelerometer data acquisition computer. A final post-deployment status shape measurement was performed with the Vicon system with the same procedure as used for the initial shape measurement. The coordinate frame and the notation for the Vicon markers and accelerometer sensors are defined in Fig. 4.29.

Deployment testing was carried out to completion as planned without unexpected or significant issues. All three deployments behaved similarly, and they exhibited a steady initial axial acceleration, a slight asymmetric rebound when the structure reached full deployment, and full deployment after the momentary rebound. No significant axial rotation was observed, only a cantilever displacement during initial deployment, where the x+ side of the tip-interface ring was slightly ahead of the other side by about a centimeter. The rebound would typically buckle the set of tape springs nearest the base, turn the tip, and recompress the base bay, almost completely, before redeploying and locking out in its final state.

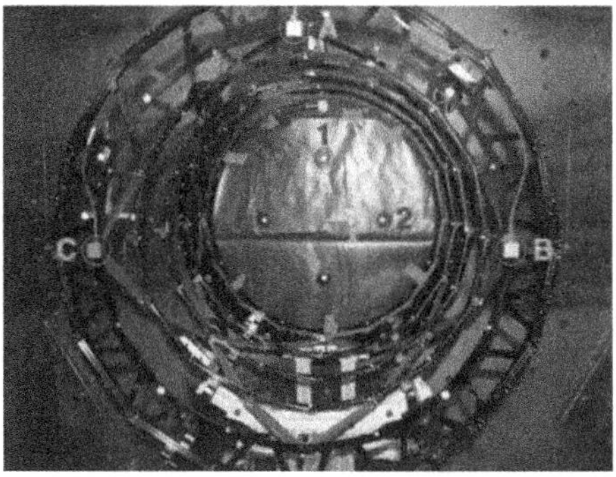

Fig. 4.29 Vicon marker and accelerometer locations for the testing of a deployable sparse isogrid column [44]. (Credit: AFRL.)

To observe peak deployment accelerations, recorded accelerometer data showed generally noisy data with high frequency content, particularly after the first bays begin to lock out at 0.25 sec and before the full deployment impact and rebound at 0.42 sec. This result implies that a bay lockout of the intermediate rings is likely exciting their local modes and transmitting the high-frequency vibrations to the tip accelerometers. This high-frequency noise in the raw accelerometer data makes it difficult to extract and observe the impulse at the end of free-deployment and when the structure rebounds. In addition, the measured acceleration intermittently exceeds the maximum 50 g range and once at the rebound point with a 51.6 g reading. It is difficult to discern whether the impact has an additive high-frequency component or if the measurement is real, since it is beyond the sensor capability. For a true measurement, the experiment needs to be performed again with accelerometers with a greater acceleration range.

However, the intent of the experiment is to gauge the impulse energy from free-deployment that would be negatively imparted onto a spacecraft bus. Typically, high-frequency vibrations are low energy, quickly damp out, or may be easily handled by dampers. The low-frequency, high-energy vibrations are of particular interest. The accelerometer data filtered by a fifth-order Butterworth low-pass filter with a 30 Hz cutoff frequency showed the major deployment events of release, initial bay lockout, rebound impulse, lockout at deployment completion, and settling. The Vicon calculated acceleration and accelerometer data were compared and found to be in good agreement. Characterizing the free-release deployment behavior of space structures has not been straightforward until the recent availability of videogrammetry metrology systems developed for the entertainment industry. The Vicon videogrammetry system successfully captured all relevant acceleration and kinematic behaviors throughout the rapid deployment duration, showing strong correlation to accelerometer measurements and high-speed photography frames.

REFERENCES

[1] AIAA Standard—Moving Mechanical Assemblies for Space and Launch Vehicles (S-114-2005).

[2] NASA General Environmental Verification Standard (GEVS), NASA-STD-7000A, 2013.

[3] Adams, D. S., and Mobrem, M., "MARSIS Antenna Flight Deployment Anomaly and Resolution," AIAA 2006-1684, 2006.

[4] Robinson, S. A., "Simplified Spacecraft Antenna Reflector for Stowage in Confined Envelopes," U.S. Patent 5,574,472, 1996.

[5] Japan Aerospace Exploration Agency Safety and Mission Assurance Department, http://sma.jaxa.jp/en/index.html, last accessed April 11 2017.

[6] Meguro, A., and Mistugi, J., "Ground Verification of Deployment Dynamics of Large Deployable Space Structures," *AIAA J. of Spacecraft and Rockets*, Vol. 29, No. 6, 1992.

[7] Sawada, H., Mori, O., Okuizumi, N., Shirasawa, Y., Miyazaki, Y., Natori, M., Matunaga, S., Furuya, H., and Sakamoto, H., "Mission Report on The Solar Power Sail Deployment Demonstration of IKAROS," 52nd AIAA/ASME/ASCE/AHS/ASC Structures, Structural Dynamics and Materials Conference, AIAA-2011-1887, Denver, CO, 2011.

[8] Meguro, A., Ishikawa, H., and Tsujihata, A., "A Study on the Accuracy of Deployment Testing for Large Deployable Structures," 43rd AIAA/ASME/ASCE/AHS/ASC Structures, Structural Dynamics and Materials Conference, AIAA-2002-1606, Denver, CO, 2002.

[9] Ozawa, S., Shintate, K., and Tsujihata, A., "Tri-Fold Deployable Reflector for Communication Satellites," 29th AIAA International Communications Satellite Systems Conference, AIAA-2011-8022, Nara, Japan, 2011.

[10] Shirasawa, Y. et al., "Evaluation of Membrane Dynamics of IKAROS Based on Flight Result and Simulation using Multi-Particle Model," *Trans. JSASS Aerospace Tech. Japan*, Vol. 10, No. 28, 2012.

[11] Shirasawa, Y. et al., "A Study on Membrane Dynamics and Deformation of Solar Power Sail Demonstrator IKAROS," 53rd AIAA/ASME/ASCE/AHS/ASC Structures, Structural Dynamics, and Materials Conference, AIAA 2012-1747, 2012.

[12] Meguro, A., Shintate, K., Usui, M., and Tsujihata, A., "In-orbit Deployment Characteristics of Large Deployable Antenna Reflector onboard Engineering Test Satellite VIII," 58th International Astronautical Congress, IAC-07-C2.2.02, Hyderabad, 2007.

[13] Ozawa, S., Fujiwara, Y., and Tsujihata, A., "Development of corotational formulated FEM for application to 30 m class large deployable reflector", *IOP Conference Series: Materials Science and Engineering*, Vol. 10, No. 1, 2010.

[14] Uchimal, et al., "Development of Large Deployable Reflector: Surface Accuracy," *The Institute of Image Information and Television Engineers Technical Report*, Vol. 25, No. 73, pp. 57–64, 2001.

[15] Sawada, H. et al., "Shape Estimation of IKAROS's Solar Power Sail by Images of Monitor Cameras," 53rd AIAA/ASME/ASCE/AHS/ASC Structures, Structural Dynamics, and Materials Conference, AIAA 2012-1749.

[16] Tsuda, Y. et al., "Achievement of IKAROS — Japanese Deep Space Solar Sail," *Acta Astronautica*, Vol. 82, 2013, pp. 183–188.

[17] Tsuda, Y. et al., "Flight Status of IKAROS Deep Space Solar Sail Demonstrator," *Acta Astronautica*, Vol. 69, 2011, pp. 833–840.

[18] Miyazaki, Y. et al., "Wrinkle/slack Model and Finite Element Dynamics of Membrane," *International Journal for Numerical Methods in Engineering*, Vol. 66, 2006, pp. 1179 – 1209.

[19] Shirasawa, Y., Mori, O., Miyazaki, Y., Sakamoto, H., Hasome, M., Okuizumi, N., Sawada, H., Furuya, H., Matsunaga, S., Natori, M., and Kawaguchi, J., "Analysis of Membrane Dynamics using Multi-Particle Model for Solar Sail Demonstrator IKAROS," 52nd AIAA/ASME/ASCE/AHS/ASC Structures, Structural Dynamics and Materials Conference, AIAA-2011-1890, Denver, CO, 2011.

[20] Tsuda, Y. et al., "Flight Result and Analysis of Solar Sail Deployment Experiment Using S-310 Sounding Rocket," *Space Technol.*, Vol. 26 (1 – 2), 2006, pp. 33 – 39.

[21] Tsuda, Y. et al., "Microsatellite-Class Solar Sail Demonstrator – Mission Design and Development Status," 25th International Symposium on Space Technology and Science, 2006-k-37, 2006.

[22] Mori, O et al., "Static Deployment of Large Membrane of Spinning Solar Sail Using a Balloon," 17th AAS/AIAA Space Flight Mechanics Meeting, AAS07-67, 2007.

[23] Sakamoto, H. et al. "Transient Dynamic Analysis of Gossamer-Appendage Deployment Using Nonlinear Finite Element Method," *Journal of Spacecraft and Rockets*, Vol. 48, No. 5, Sept. – Oct. 2011.

[24] Miyazaki, Y. et al., "Conserving Finite Element Dynamics of Gossamer Structure and Its Application to Spinning Solar Sail IKAROS," 52nd AIAA/ASME/ASCE/AHS/ASC Structures, Structural Dynamics, and Material Conference, AIAA 2011-2181, 2011.

[25] Shintaku, K. et al., "Experiment and Analysis of Spinning Membrane Deployment Focusing on Shape Imbalance for Solar Sail," *Trans. JSASS Aerospace Tech. Japan*, Vol. 12, No. 29, pp. Pc_55 – Pc_62, 2014.

[26] Satou, Y. et al., "Nonflatness of Solar Sail Membrane Predicted by Nonlinear Finite Element Analysis," 1st Spacecraft Structures Conference, AIAA SciTech Forum, AIAA 2014-0832, 2014.

[27] Wright, D., "Design, Integration, and Testing of an Advanced Synthetic Aperture Radar," *IUTAM-IASS Symposium on Deployable Structures: Theory and Applications*, S. Pellegrino, and S.D. Guest (Eds.), Kluwer Academic Publishers, 2000.

[28] Murphey, T. W. et al., "High Strain Composites," Proceedings of the 2nd AIAA Spacecraft Structures Conference, Kissimmee, FL, 2015.

[29] Turse, D. et al., "Flight Testing of a Low Cost De-Orbiting Device for Small Satellites," Proceedings of the 42nd Aerospace Mechanisms Symposium, Baltimore, MD, 2014.

[30] Okuizumi, N. et al. "Small-scale Experiments and Simulations of Centrifugal Membrane Deployment of Solar Sail Craft 'IKAROS,'" 52nd AIAA/ASME/ASCE/AHS/ASC Structures, Structural Dynamics and Materials Conference, AIAA-2011-1890, Denver, CO, 2011.

[31] Satou, Y. et al., "Deformation Properties of Solar Sail IKAROS Membrane with Nonlinear Finite Element Analyses," 2nd AIAA Spacecraft Structures Conference, 2015.

[32] Murphy, D., "Validation of a Scalable Solar Sailcraft System," *J. of Spacecraft and Rockets*, Vol. 44, No. 4, 2007.

[33] NASA Plum Brook Station Facilities, https://www.nasa.gov/centers/glenn/about/testfacilities/index.html, Feb. 2, 2017.

[34] Shing, P. B., "Real-Time Hybrid Testing Techniques," *Modern Testing Techniques of Structural Systems*, O.S. Bursi, and D. Wagg (Eds), CISM International Centre for Mechanical Sciences, Vol. 502, pp. 259–292, 2008.

[35] Stoten, D.P., Tu, J-Y., and Li, G., "Synthesis and control of generalized dynamically substructured systems," Proceedings of the Institution of Mechanical Engineers, Part I: Journal of Systems and Control Engineering, 2008.

[36] McEachen, M. E., and Truatt, T.A., "Confirmation of New Analytics for Ultra-Light Lattice Column Strength Using a 40-m Flight Article," 50th AIAA/ASME/ASCE/AHS/ASC Structures, Structural Dynamics & Materials Conference, AIAA-2009-2169, Palm Springs, CA, 2009.

[37] Footdale, J. N., and Murphey, T.W., "Mechanism Design and Testing of a Self-Deploying Structure Using Flexible Composite Tape Springs," Proceedings of the 42nd Aerospace Mechanisms Symposium, Baltimore, MD, 2014.

[38] Penn, J., Johnson, C., Lewis, J., Dear, T., and Stewart, A., "GPM Solar Array Gravity Negated Testing," Proceedings of the 42nd Aerospace Mechanisms Symposium, Baltimore, MD, 2014.

[39] Footdale, J. N., Banik, J. A., and Sanders, J. W., "Design and Deployment Testing of the Multi-Arm Radial Composite (MARCO) Reflector Antenna," Proceedings of the 3rd AIAA Spacecraft Structures Conference, San Diego, CA, 2016.

[40] Silver, M. J., Hinkle, J. D., and Peterson, L. D., "Controlled Displacement Snap-Through of Tape Springs: Modeling and Experiment," 46th AIAA/ASME/ASCE/AHS/ASC Structures, Structural Dynamics and Materials Conference, AIAA-2005-2314, Austin, Texas, 2005.

[41] Johnson, C. D., Wilke, P. S., and Pendleton, S. C., "SoftRide Vibration and Shock Isolation Systems that Protect Spacecraft from Launch Dynamic Environment," Proceedings of the 38th Aerospace Mechanisms Symposium, Langley Research Center, 2006.

[42] Giuseppe, G., and Brown, J., "Developments in Non-Contact Measurement Using Videogrammetry," www.geodetic.com.

[43] Gruen, A. "Fundamentals of Videogrammetry—A Review," *Human Movement Science*, Vol. 16, pp. 155–187, 1997.

[44] Jeon, S. K., Banik, J. A., and Peterson, M. E., "Free Deployment of a Sparse-Isogrid Column with High Strain Composite Ribs," 3rd AIAA Spacecraft Structures Conference, 2016.

[45] Ardelean, E.V., Gooding, J.C., Coombs, D.M., Griffee, J.C., Babuska, V., Robertson, L.M., and Lane, S.A., "Cable Effects Study: Tangents, Rat Holes, Dead Ends, and Valuable Results," 51st AIAA/ASME/ASCE/AHS/ASC Structures, Structural Dynamics and Materials Conference, AIAA-2010-2806, Orlando, FL, 2010.

[46] Robertson, L. et al., "Cable Effects on The Dynamics of Large Precision Structures," 48th AIAA/ASME/ASCE/AHS/ASC Structures, Structural Dynamics and Materials Conference, AIAA-2007-2389, Honolulu, HI, 2007.

CHAPTER 5

Deployment Repeatability

Olive Stohlman
NASA Langley, Hampton, VA

Mark Silver
MIT Lincoln Laboratory, Lexington, MA

Dave Waller
Ball Aerospace, Boulder, CO

NOMENCLATURE

a, b, c, d	Actual lengths of truss members
a_0, b_0, c_0, d_0	Exact design lengths of members
e	Distance between base support connection points
x, y, θ	Errors in the deployed shape of a bay
$x_{\text{boom}}, y_{\text{boom}}, \theta_{\text{boom}}$	Errors in the deployed shape of a boom
N	The number of bays in a boom
σ_a	The standard deviation of values of a

5.1 INTRODUCTION

Each time a large ultra-lightweight spacecraft (LULS) structure is deployed in space, it will have a slightly different deployed shape. This variation arises from small changes in the hardware and environment, like migration of joint lubricants, changes in the sun angle, and other factors that cannot be perfectly repeated from one deployment to the next. What distinguishes LULS from conventional spacecraft is the scale and complexity of their deployments and the often-small margin between shape performance and requirements.

The goal of this chapter is to help the reader think about deployment repeatability: to assess project needs before planning tests, to identify likely factors in repeatability, and to design tests and models. This process is illustrated at its highest level in Fig. 5.1. First, the elements of shape error should be assessed, so that a verification plan can be built. This plan may rely heavily on direct measurements of the full deployment structure or on a model that takes component measurements as input. With the results of tests and models, the project may make informed decisions.

Before discussing the content of the chapter, some important terms for the discussion of repeatability need to be defined and refined.

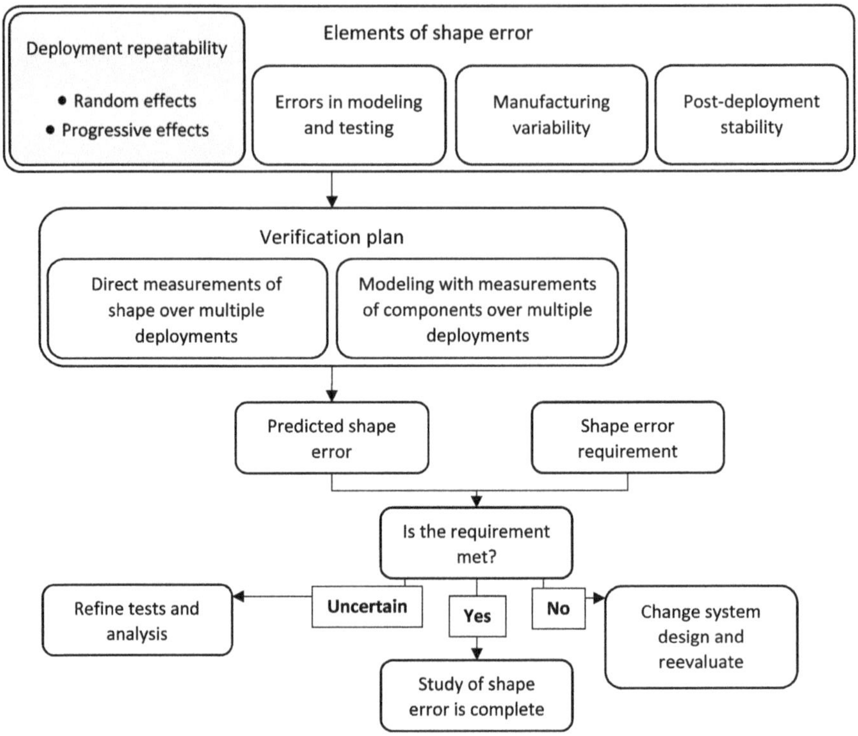

Fig. 5.1 Flowchart of shape error analysis.

For the purposes of this chapter, *repeatability* is synonymous with *precision*, a concept that is contrasted with accuracy in Fig. 5.2. Lake et al. [1] give a simple definition of deployment precision: "the error in the final deployed shape of a structure as compared to its ground-measured shape." This is the basis for the definition of repeatability, but a few adjustments must be made in the context of LULS structures. First, large ultra-lightweight structures are not necessarily expected to match a ground-measured shape. Instead, they are expected to match some predicted shape that may combine ground measurements and analysis; the analysis introduces a factor of uncertainty. Figure 5.3 shows this distinction, where a known shift between ground and space conditions can be offset by design, but a lack of precision cannot. Second, it needs to be clear that repeatability is not about a single final deployed position; rather, repeatability concerns *the size of the envelope of statistically likely deployed shapes.*

Section 5.2 summarizes the most important lessons of this chapter. In particular, it discusses the preference for full-structure testing over substructure testing, some high-level categories to help engineers think about the nature of the shape variability they see in testing, and the high-level sources of deployed shape

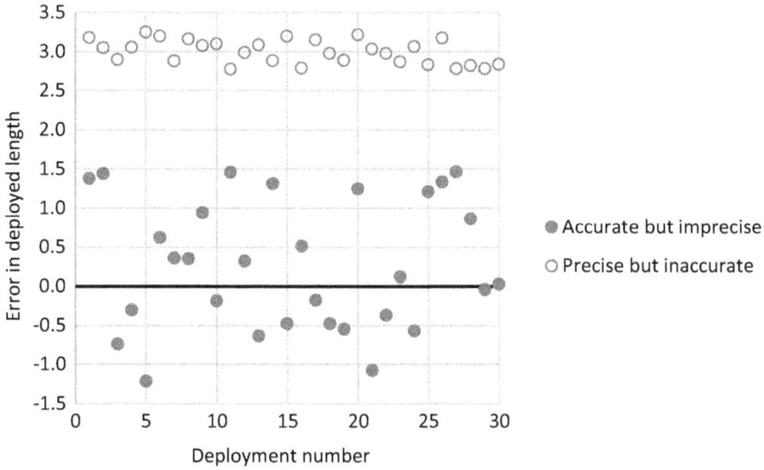

Fig. 5.2 The distinction between accurate and precise booms over many deployments, based on two-dimensional tip position.

variability. Section 5.2.3 is a step-by-step list for approaching deployment repeatability in a project with requirements. Least-precision LULS structures may find, by following the steps described in Section 5.2.3, that there is no need for a detailed study of repeatability at all. Deployment repeatability is not equally important to all projects, and throughout this chapter there are reminders to avoid unnecessary work.

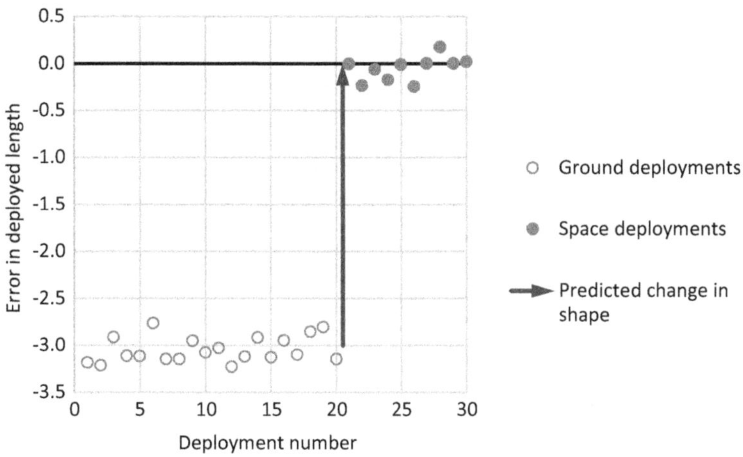

Fig. 5.3 Repeatability in a structure that is accurately modeled.

At a high level, repeatability testing is broken into two areas of practice: testing of complex structures and testing of components for use in models. Section 5.3 discusses complex structure testing. In this type of testing, much of this book is relevant, and deployment repeatability testing may simply be repeated deployment testing. Section 5.3.1 reviews some references on structural settling as a topic distinct from repeatability. Section 5.3.2 discusses the repeatability of test fixtures and equipment, and Section 5.3.3 presents a recommended step-by-step approach to testing large ultra-lightweight structures for repeatability.

Component-level repeatability testing, the subject of Section 5.4, might be pursued as its own end for products or as an input to a model. This level of testing, which is discussed in an overview in Section 5.4.1, is common for joints. Section 5.4.2 includes a worked example of how the statistics of error propagation can be applied to repeatability for moderate-precision applications (as defined in Chapter 2).

Section 5.5 features a few case studies in repeatability testing. Section 5.5.1 details the repeatability testing of the extremely precise three-dimensional Membrane Optical Imager Real-time Exploitation (MOIRE) test article. This case study is presented as an example of state-of-the-art precision in the measurement of repeatability and as an example of practical demand for high shape accuracy. The SAILMAST and GEMS masts are discussed more briefly in Section 5.5.2. Section 5.5.3 summarizes selected case studies in repeatability (shown in Table 5.3) and a number of additional references that may be of interest to the reader.

This chapter concludes in Section 5.6 with a philosophical note about the trade between repeatability testing and complexity in shape control.

5.2 FUNDAMENTALS OF DEPLOYMENT REPEATABILITY ANALYSIS

What is the ideal way of characterizing repeatability? The more directly the test can replicate in-space deployment, the better. However, with LULS structures, it is often impossible to do a direct test without interference from gravity offload systems or the ground environment. Modeling is suitable for well-characterized parts, and stochastic modeling techniques can be used for sensitivity analysis and for generating a large sample set of trials to spot unusual cases. However, deployment repeatability is inherently a nonlinear phenomenon, which makes modeling from first principles difficult and test data a crucial input to the model.

In order of preference, the following may be considered for establishing repeatability:

1. Test every completely assembled article through full deployment as many times as statistically required.
    - Test-to-test variability in the gravity compensation is a problem unique to LULS structures. Depending on the level of repeatability required, characterization of the nonlinearities of the gravity compensation system may be necessary.

- Ideally, tests should be done in a relevant environment; if not, modeling and component testing may still be necessary to determine the significance of changes in, for example, friction.
- Testing every individual article removes the influence of manufacturing variability. When a design is single use, or when a flight article is required to be pristine and untested, it is acceptable to test an engineering model and bound the manufacturing variability.

If full deployment testing is not possible,

2. Test a statistically useful sample of parts and run a Monte Carlo analysis of a model.
    - In a repeating structure with a series of nominally identical parts (for example, a mast with a series of identical bays), an estimate can be made by combining the tested repeatabilities of the parts (see Section 5.4.2).

If parts testing or computer time for a Monte Carlo analysis are not possible,

3. Establish conservative limits of factors affecting repeatability and run models of the most extreme cases to identify worst-case outcomes.

5.2.1 SOURCES OF DIMENSIONAL ERROR

What contributes to the repeatability of the deployed shape of a structure? Random effects that vary unpredictably between deployments, and progressive effects that change predictably from one deployment to the next. Some effects can reasonably be considered random in one project, but not in another.

When one is considering sources of deployment repeatability, it is useful to look at analogous sources of errors in measurement instruments. Examples of measurement instrument error categories are described in Ref. [2]. Using measurement instruments as a foundation, deployment repeatability error sources for a single test article can be grouped into four categories.

5.2.1.1 RANDOM ERROR

Random error, shown in Fig. 5.4, is variation in the response curve throughout all tests.

Sources of random error include:

- Internal repeatability of motors or deployment drivers
- Deployment drivers
- Gravity offload systems
- Other test fixtures

Sources of random error for jointed structures include

- Free-play and friction in joints
- Loss or redistribution of lubrication

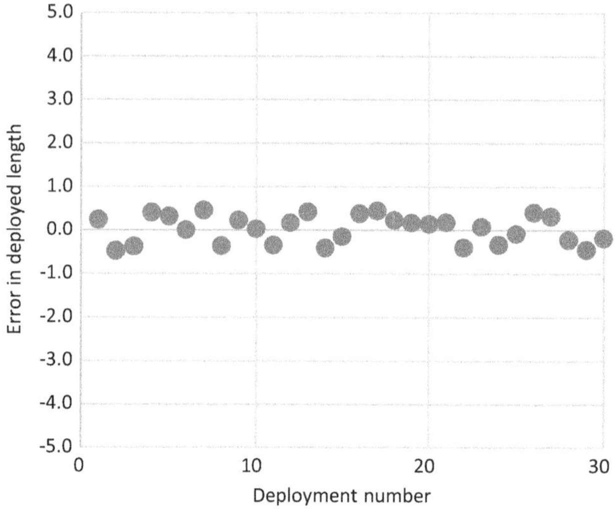

Fig. 5.4 A series of measurements showing pure random error. No known pattern relates one deployment to another.

5.2.1.2 ZERO SHIFT

Zero shift is a test-to-test shift in the zero point of the expected response curve. For the purposes of this chapter, zero shift refers to permanent changes in the structure, while reversible changes are considered random. Figure 5.5 shows a series of zero-shift errors that might occur due to repeated slips in a bolted connection.

Sources of zero-shift errors include

- Frictional slip in bolted joints within the structure during test or launch
- Slip at mounting points in the support equipment
- Material yield or failure during a test, including damage to composite fibers
- Part replacement after testing
- Joint surface wear

Deployment repeatability can also change from test article to test article, and this is known as *unit-to-unit precision error*. Sources of unit-to-unit precision error include

- Variations in materials
- Manufacturing procedures
- Manufacturing environment

Depending on whether each new article is individually tested, this can be thought of as a form of zero-shift error.

5.2.1.3 HYSTERESIS ERROR

Hysteresis errors are changes in the deployed shape based on test cycling. Figure 5.6 shows a hysteresis error that increases with every deployment.

Sources of hysteresis error include

- Test-to-test variation in the extent of the structure's deployment
- Observed trends with unidentified physical causes and dependence on the number of test cycles
- Range of possible storage time and conditions on the ground
- Range of possible storage time in space before deployment
- Plastic deformation of springs
- Material creep and relaxation

For inflatables and membranes, sources of hysteresis error include

- Viscous creep in glue
- Wrinkling
- Wear from repackaging

Note that, with LULS structures, it is understood that only a stiff structure or one measured in an excellent gravity offload system will have the same deployed shape

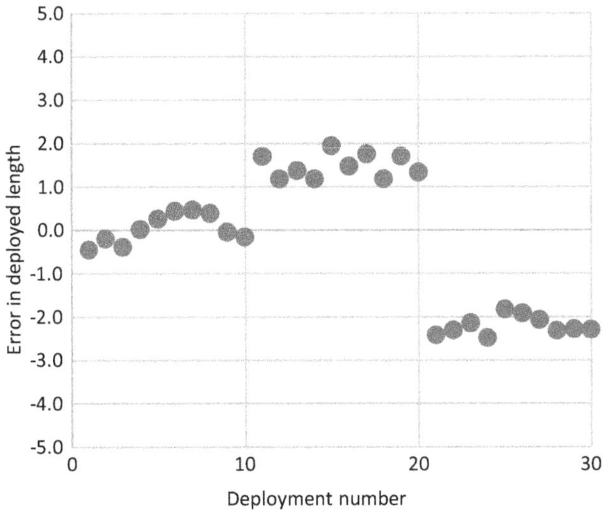

Fig. 5.5 A series of measurements showing zero-shift errors and a smaller random error. Some event has caused a permanent change in deployed shape after deployment 10 and again after deployment 20.

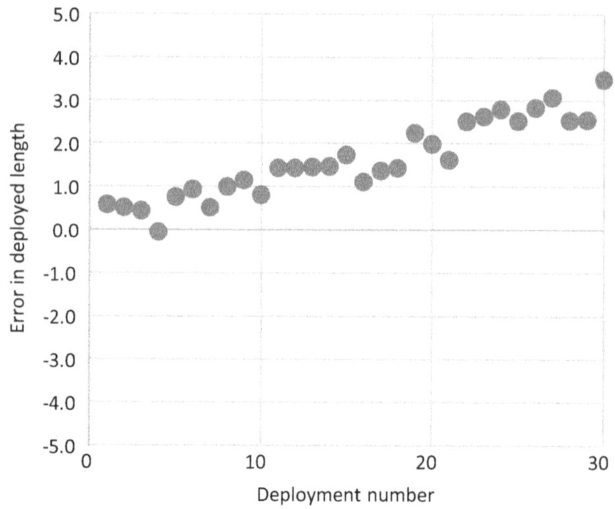

Fig. 5.6 A series of measurements showing hysteresis and a smaller random error. This trend in deployed shape depends on time or the number of deployments.

in space as it did on the ground. A gravity-related change in shape should be expected and modeled (or compensated for in testing), and generally such a change in shape is not considered to be a part of repeatability. Why, then, does this chapter include material hysteresis, which can also be modeled and is not typically random? If a particular project has well-characterized hysteresis effects, and the system design has already accommodated these effects, they sit alongside gravitational forces as part of the expected shape. If a project has uncertainty—if hysteresis might be small enough to ignore, or if it depends on factors outside the project's control (like time in storage, or mounting position in the launch vehicle)—then it should be placed alongside other repeatability factors.

5.2.1.4 ENVIRONMENTAL SENSITIVITY ERRORS

Environmental sensitivity errors are variations in the deployed shape due to changes in the environment. Extremely well-characterized and well-controlled large ultra-lightweight satellites will be able to predict some environmental effects as part of a model, while very basic large ultra-lightweight satellites may have to treat them as random. In an advanced satellite, the spacecraft attitude may always be controlled to great precision, but in a CubeSat, there may be no attitude determination at all. Such a CubeSat might treat sun angle and tumbling rates as randomly distributed, and it will consequently have a randomly distributed range of thermal deformations. Thus, the advanced satellite has a modeling challenge and the CubeSat has an increased uncertainty of shape error. Figure 5.7 shows two patterns that could be attributed to environmental errors. In each pattern, there is a sudden change in performance due to a change in environment.

Sources of environmental error include

- Randomness in spin rate and spacecraft dynamics (where these affect the deployment specifically, for example by biasing the initial position of a joint)
- Randomness in the environment, for example where the attitude is not controlled or where friction changes unreliably in vacuum or thermal conditions (where these affect the deployment, and not only post-deployment stability)
- Temperature
- Humidity, pressure, and airflow changes affecting ground testing

Many of these more obviously affect post-deployment stability, but some can alter the deployment process itself.

Establishing worst-case limits or a statistical range for each of these effects is a good way to bound repeatability concerns. Occasionally returning to the lists of error sources in this section may reveal that repeatability is well controlled and that no further testing is needed.

Many of these same effects can be found in the test equipment that is used to evaluate repeatability. Repeatability in the test fixtures themselves is addressed in Section 5.3.2, later in this chapter.

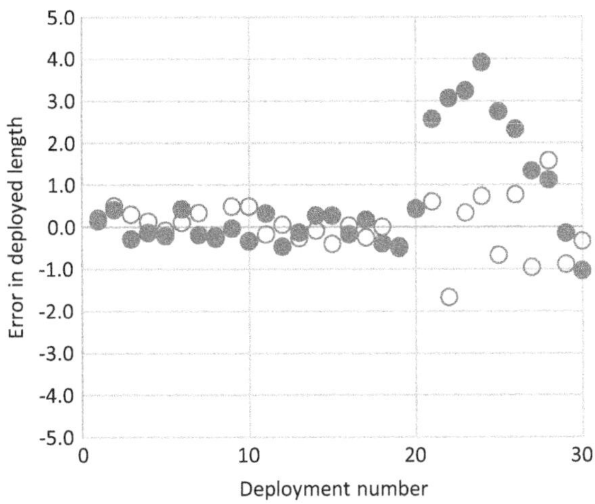

Fig. 5.7 Two series of measurements showing environmental error after an environmental change at deployment number 20. Environmental error could change the amount of random error, add a zero shift, or introduce a hysteretic trend.

5.2.2 SELECTION OF TEST VARIABLES

What shape measures are important to the mission? It is important to consider what measurements would provide confidence in deployment repeatability. Ideally, the test variables should be inspired by relevant requirements or modifiable design parameters. Consider the example of NuSTAR [3], an x-ray telescope with a 10 m mast separating its focusing mirrors from its detectors. The aim of the mirrors on the detectors was particularly important, and the telescope was relatively insensitive to errors in boom length. Three degrees of freedom (DOF) might be sufficient to describe the envelope of deployment repeatability in this case, describing the position and rotation of the focused image. For a case like this, an engineering model of the flight metrology system, which measures these values, could work well for testing.

In another example, the ideal measurement of a deployed parabolic radio frequency (RF) reflector system would be a functional RF range measurement. However, this measurement would require a large RF testing range, which may be unfeasible, or at least unfeasible for a series of deployment repeatability tests. When the structure cannot be measured as it will be used, a secondary measurement that then needs to be related to the functional target must be made. For a membrane RF reflector, a secondary measurement could be dot projection photogrammetry [4]. Or for RF mesh reflectors, Digital Image Correlation (DIC) with distributed targets may be necessary. As the spatial resolution of the secondary measurement decreases, larger analytical leaps are needed to translate the lower-resolution repeatability measurement into an estimate of functional repeatability.

5.2.3 STEP-BY-STEP GUIDE TO REPEATABILITY ANALYSIS

At the highest level, repeatability analysis starts with assessing the project's needs. Diving directly into experimental design can lead to excess work and tests that a poorly suited to real engineering needs. Before designing a repeatability test, consider these steps for a workflow that is informed by project goals.

1. Establish a limited number of figures of merit for deployed shape; ideally, these should be directly related to project requirements with as few analytical steps as possible.
2. Gather existing data: What deployment data is available? What models have been built? Compare existing data with the checklist of factors in repeatability (Section 5.2.1), review the structure for factors to add to the checklist, and establish a range of values for each factor based on the best existing understanding.
3. Establish worst-case limits on repeatability based on existing data and models; if this is acceptable, end the study of repeatability.
4. Design a study to establish repeatability, giving preference to the types of testing and analysis outlined at the beginning of Section 5.2.

5. Collect data.
6. Evaluate the repeatability envelope and shape error envelope, and compare these with shape error requirements. Ask these questions: Is the requirement met? Is redesign necessary to meet the requirement? Is further testing required to answer these questions?

There is an inherent trade between confidence in deployment repeatability and the number of deployments sampled, and deployment testing may be costly to repeat. In order to quantify the effect of this trade, a good statistical starting point is Student's t-distribution and the construction of prediction intervals for a population with unknown mean and variance, a subject not covered in this book.

5.3 FULL STRUCTURE AND COMPLEX SUBSTRUCTURE TESTING

Measurements of an entire LULS can be directly connected to performance, without intermediate modeling or allowances for interfaces. This advantage justifies some additional expense and a great deal of additional care in the test setup.

5.3.1 STRUCTURAL SETTLING DIMENSIONAL ERROR

After a structure deploys, its shape will still change over time. The key question is whether this change is large enough to affect system performance. Therefore, it is desirable to characterize a structure both immediately after deployment and after some settling time. The structure may actually have better repeatability after some settling time than the earlier measurement suggests, due to two effects: microlurching and material creep.

A structure with sliding joints may be subject to post-deployment *microlurching*, where the strain energy of the structure is relieved by a series of frictional stick-slip events in the joints [5]. Lake et al. [1] describe the effect well:

> [. . .] it appears that structures tend to change their shape slightly after deployment to relax internal stresses built up by the deployment process. After structures have undergone a post-deployment, progressive microlurch, subsequent microlurching responses tend to be random but bounded in magnitude. The boundary of this random microlurch response is the equilibrium zone of the structure.

Thermal snap (or thermal creak) [6, 7] is one source of microlurching; microlurching can also be encouraged by intentionally exciting the structure. This early microlurching towards strain relief should be distinguished from later microlurching that affects post-deployment stability, long after deployment.

Material creep is particularly likely to be a factor when the structure has been stowed in a substantially different state of stress from its deployed condition. At one extreme, a tensioned truss structure goes from low stress when stowed to high stress after deployment. At the other extreme, a strain-energy-deployed tape spring starts in a condition of high material stress and deploys to a low-stress shape. In either case, the deployed shape can vary with stowage time and the settling time allowed after deployment.

In order to accommodate settling effects, several measurements of the deployed shape should be taken after a single deployment. A settling time, after which further motion is below the threshold of the experiment, can be identified from these data. Final repeatability measurements should be taken after this fixed settling time.

5.3.2 CHARACTERIZATION OF REPEATABILITY IN TEST EQUIPMENT

In order to have a high level of confidence in ground-based repeatability measurements, the overall repeatability inherent in the test equipment and facility should be considered. For this discussion, test equipment includes

- Test article mounting fixtures
- Offload mechanisms
- Offload mounting fixtures
- Sensors and associated data acquisition and storage equipment
- Sensor mounting fixtures
- Structural connections between all mounting fixtures

The only aspects of test equipment repeatability unique to or especially challenging to large ultra-lightweight structures are those related to the offloading system and the structural connections between all mounting fixtures. Offload systems are usually designed for low mass, low friction, and compliance, with little concern for repeatability. For smaller structures, a support frame can be used to connect the test article, sensors, and other fixtures into one relatively rigid unit. For the structures considered in this book, this often cannot be achieved. Therefore, characterizing the repeatability of all test equipment is important when quantifying the test article's repeatability.

In addition to the test equipment, test-to-test changes in the testing facilities may need to be considered. This can include environmental effects such as temperature, humidity, pressure, and air flow in addition to motion of the floor, doors being slammed, or other dynamic events. Concern about facility effects on measurements is not unique to the testing of LULS structures, but their effects may show up in different ways. A key example is the cyclic airflow in the heating or air conditioning system pushing on the structure, causing deformation during one measurement but not the next. Testing in a closed chamber, directing the air away from the test or isolating the test area with walls or curtains, can abate this. Examples of the use of test chambers or curtains to limit airflow in large space structure measurements can be found in Refs. [8, 9].

In order to minimize test equipment repeatability error, some forethought should be given when designing the test. First of all, consider the room where the testing will take place. Often, the size of the structure being tested dictates which room will be used, so a test engineer does not have much choice in the matter. Things to note when considering a room are

- What other activities are happening in the room, in neighboring rooms, and in the building?
- Who else has access to the room, and when will they be there?
- If you are sharing the space, will you have to move your equipment?
- How is the room heated and cooled?
- Does the room have humidity control?

It is a good idea to make long-duration measurements of the thermal, humidity, and vibration conditions in the room to quantify the variations that will be seen ahead of time. The measured variations can then be compared with thermal expansion or moisture expansion data for the materials in the test article, and vibration levels can be used in analytical models of the test article to see if the induced shape variations are of concern for the repeatability levels desired. Periodic changes throughout the day in temperature or humidity may reflect the cycle of the heating or cooling air handler, and airflow in the room can be further investigated. When one is characterizing the vibration environment, low-frequency vibration measurements should be made in the room wherever test equipment will be mounted. For an overhead offload system, this may include the ceiling. If the levels and frequency of vibrations measured are of concern, some form of isolation should be used.

Once the environment in the room is characterized, the test fixtures can be designed to minimize the room's environmental effects. As mentioned previously, in the testing of large structures, the mounting fixtures for the test article, sensors, and offload may be distributed around the room so that the building acts as the structural connection between all mounting fixtures. This can be problematic for repeatability tests because some or all of these fixtures may not be well constrained to the building. The fixtures could be inadvertently moved between tests, resulting in a zero shift in the repeatability measurements. This can be addressed in two ways:

1. The fixtures can all be mounted with sufficient stability so as to not affect measurements.
2. Measurements of the test article and fixtures can be made with respect to fixed references within the room.

The first solution is self-explanatory, but depending on the level of stability required, it may be difficult to achieve in practice. The second solution requires regular reference measurements, with a metrology system capturing the relative location of the test article and mounting fixtures to the fixed references. Then, any relative motion can be removed from the repeatability measurements during data post-processing.

Because the offload system moves in parallel with the structure, the repeatability of the offload system alone can be difficult to quantify. For a mechanical offload system, measurements of the locations of the offload points in the deployed state with some form of metrology system can be useful to verify that similar extents of offload motion are reached from test to test. In all offload systems, it is useful to verify that consistent offload forces are being applied from test to test. For off-loading balloons this can be as simple as measuring the lift force between tests. For a mechanical system, this may mean checking the spring stiffness of each offload connection or even using force sensors in line with each offload connection.

5.3.3 FULL STRUCTURE REPEATABILITY TESTING PROCESS

Repeated tests of the complete, assembled structure are the ideal method of determining deployment repeatability. They provide the most convincing evidence of performance and eliminate many sources of uncertainty. Because high-precision results can be obtained with this type of testing, it is particularly rewarding to plan tests well, following every step of this process.

5.3.3.1 DETERMINE THE TEST MATRIX

The testing process begins with determining what repeatability means for the test article and what tests need to be performed, as described in Steps 1 through 4 of the step-by-step guide to repeatability analysis, provided in Section 5.2.3, earlier in this chapter. When this is complete, the repeatability test matrix will be defined. The number of repeatability tests to be performed should be determined from the statistical confidence desired but may be limited in practice by schedule. Consider the effects of creep and stress relaxation on the test article in both the stowed and deployed states. If these material effects might be drivers for repeatability, measure repeatability for a variety of stow durations.

5.3.3.2 CHARACTERIZE THE TEST FACILITY

After the test matrix is established, the test facility must be considered. If possible, select a test facility that will provide the environmental stability needed for the measurement. Characterize the environment of the test facility. This may include characterization of vibration, temperature, humidity, and air flow.

5.3.3.3 SELECT SENSORS AND UNDERSTAND THEIR REPEATABILITY

The appropriate measurements for the desired functional repeatability were identified when the test matrix was determined. When selecting sensors to make these measurements, consider their repeatability and long-term stability in the characterized test environment. If these data are not available from the manufacturers for commercial sensors, calibration measurements within the desired environment may be required.

5.3.3.4 PERFORM CALIBRATION MEASUREMENTS OF THE ENTIRE TEST CONFIGURATION

Once the repeatability of the measurement sensors is well understood, the positional stability of all test article, sensor, and gravity offload mounting points should be considered. If any of these might be too unstable for the desired measurements, identify stable measurement references as needed so that motion of the test equipment can be removed from the final data. If possible, characterize repeatability of all test apparatus (sensors, mounting frames, gravity compensation system, and so on) within the test environment over all expected environmental conditions using a stable test reference.

5.3.3.5 PERFORM TESTS

Mount the test article to the test frame and gravity compensation system, and perform tests as determined in the test matrix. Track the times of all tests (both stowed and deployed), so that if creep occurs during or between tests, the timespan for the creep is documented. Process data as soon as possible to identify any necessary changes to the test setup.

5.4 COMPONENT-LEVEL TESTING AND MODELING

In some situations, a project might want to incorporate component-level testing with a repeatability-oriented model:

- In situations where testing the assembled structure is prohibitively expensive or difficult.
- In situations where the design can accommodate some changes as an outcome of a sensitivity study.
- In situations where part replacements might be done without retesting the integrated structure.

With excellent measurements of the constituent parts, modeling can capture repeatability effects. However, some caution is warranted, because modeling carries a risk of overlooking "unknown unknowns," leaving open the possibility of real-world effects that are not captured in the model because they stem from properties or situations that are not modeled. Some degree of this risk is already accepted if microgravity and thermal deformation are to be modeled, as is often the case in difficult-to-test LULS.

Knowing what to include in the model requires a clear understanding of what might affect repeatability. Refer to the list of factors affecting repeatability (provided in Section 5.2.1, earlier in this chapter) as a checklist for potential inclusion in the model. Some additional questions to ask include the following:

- *What can be summarized and reduced to simpler-to-model effects?* Because the model will have to be rerun many times, simple abstractions of complex parts have great advantages. This philosophy should guide testing. It is usually better to test whole hinges for hysteretic load-displacement paths and put these data into a computationally cheap 1-DOF abstraction of a hinge than to test the hinge's surface roughness and apply this to a computationally expensive solid contact model of the hinge. This is especially true in a structure with many repeating parts, and in an analysis that will be run many times for statistical purposes.

- *Is it necessary to model the deployment process, or is it sufficient to initialize the model in the fully deployed state?* Some types of path-sensitive material damage might be well captured by a fully dynamic deployment simulation. Perhaps a deployment where each joint is driven into place by a motor could be sensitive to small differences in motor controller timing. In these cases, the analyst might choose to model the entire deployment path, with knowledge of the material damage model or motor controller timing precision. On the other hand, if many repeated and environmentally representative deployment tests can be performed on the joints, it is not necessary to model the deployment process. The test data can be integrated into each joint as part of its constitutive model, initialized as a preload on the structure, or seeded as an imperfection by a variety of methods.
- *What components, if any, might be swapped out between testing and flight?* These components deserve a closer look at manufacturing variability, and it may be a simple modeling task to consider a statistical distribution of their properties.
- *Is it necessary to capture the as-built structure?* If the model is answering questions about repeatability only, a highly accurate prediction of the actual shape may be unnecessary. That is, a model with a realistic distribution of properties may be just as good as a model with the exact properties of each as-built part. For example, if one or two deployment tests are possible and show excellent accuracy, the project may be concerned only with the likelihood that subsequent deployments will deviate from those tests. In this case, where there is no interest in refining the shape prediction itself, reproducing the as-built shape is probably not important.
- *Where is there friction in the structure?* Explicitly list every source of friction before deciding whether some will be neglected.
- *What will change without the preload of gravity? In the space thermal environment? In a vacuum?* Joint friction, for example, can be sensitive to all of these.
- *What are the subjects of the trade or sensitivity study?* Subjects for an analysis trade study could include preload in tension elements, acceptance criteria for joint surface smoothness, load levels and timing control in a driven deployment, the number of acceptable previous deployment of tapespring hinges, and any number of things that are difficult to systematically vary in physical testing.

5.4.1 METHODS FOR TESTING JOINTS OR FOLDED MEMBERS

Single joint repeatability tests are often performed to verify new joint design concepts [10, 11, 12] or to produce data to update the joint characteristics in a larger model [13]. Uncertainty of the load in the joint at the end of deployment could be large enough that nonlinearity and hysteresis in the joint response will be a significant contributor to repeatability error. It is important to quantify the load range expected in the deployed joint and to perform tests over that range.

For elastically folded joints, coiled members, and tapespring-style booms, hysteresis from time spent in the stowed state is important to quantify. The structure will likely be in this stowed state for a long time, under varying environmental conditions, resulting in significant stress relaxation or creep before deployment. For elastically stowed members, gathering deployment repeatability data for different stow durations (and possibly different stowed environments) is recommended.

Testing the deployment repeatability of these relatively small components can be much easier than testing the larger assembled structure because a common fixture for both the component under test and the measurement sensors often can be used. Additionally, gravity offload may not be necessary, and the whole test setup may fit within available environmental chambers.

5.4.2 APPROXIMATE ESTIMATION

Component-level tests can be combined with a model of the full structure to make an educated estimate of a structure's deployment repeatability. This section highlights published work on simple models that capture the order of magnitude of a structure's deployment repeatability, followed by a simple worked example.

5.4.2.1 APPROXIMATE ESTIMATION IN LITERATURE

Simple assessments of deployment repeatability can be accomplished for structures where the sources of error are well understood and can all be categorized as "random error." Reference [14] describes a method that is particularly well suited to joint-dominated masts with repeating bays. It consists of a set of tests and a simplified analysis:

- Each member of a large group of joints is tested for repeatability over repeated actuations.
- The joint test results are compiled into a database of joint performance, with a characteristic standard deviation of the deployed position of each DOF in each joint.
- Another database is built of linear relationships between small errors in the joints and shape errors in the full bay, and of linear relationships between bay shape errors and mast shape errors.

With this database of characteristic joint repeatabilities and the relationship between joint errors and bay errors, it is possible to estimate the overall mast performance (emphasis added):

> *Finally, the root sum of the squares of the various distortion sources (primarily joints) in each component in the relevant degrees of freedom (typically 6) was taken to be the predicted standard deviation of the system deployment repeatability. The results of this simple yet powerful algorithm was compared against hardware test results from a wide variety of deployable systems (coilable and articulated booms, articulating hexapod structures, etc.) and found to correlate to system repeatability measurements* ***within about a factor of two****. This predictive accuracy is more than adequate for the purposes of mission planning, both in the selection of structure type, and in the allocation of alignment budgets early on in the development process* [14, p. 5].

Heald and Peterson [15] concur with the general theme that this method is appropriate only for order-of-magnitude estimation, finding that "the uncertainty predicted by applying the kinematics of the structure to the latch repeatability results exceeded the uncertainty that was actually measured by 224%," for a simple hinge and latch telescope petal deployment assembly. Based on the extremely limited published data, modeling of an assembly coupled with repeatability measurements of its joints could be reasonably expected to predict a repeatability in the range of 20 percent to 500 percent of the true assembly repeatability, if the assembly is joint dominated and well suited to this type of modeling.

This method is particularly cost effective when the same joint design will be used many times in one structure, or across many structures.

Simply having an order-of-magnitude estimate of a mast's repeatability may show that it is repeatable enough to skip multiple deployment tests or may show that the structure is certainly not acceptable and will require a shape adjustment mechanism. If neither case applies, a more detailed assessment is called for.

5.4.2.2 WORKED EXAMPLE OF REPEATABILITY PROPAGATION

This section contains a simple worked example of repeatability propagation, which follows the same mathematical principles as any error propagation. Another worked example, for a geometrically simple deployable telescope petal, can be found in Ref. [16].

Consider a two-dimensional, statically determinate mast bay, shown in Fig. 5.8. Every bay is intended to be identical. The deployment and latching system is a part of the diagonal member, so the diagonal length snaps into place with some variability. The longerons are pin jointed at either end and have some functional length variability.

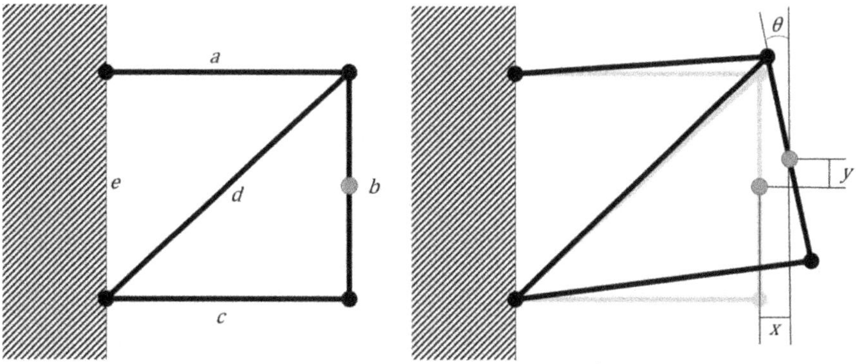

Fig. 5.8 Variables of the 2-D example mast bay.

The nominal lengths of the bars are, for this example,

$$a_0 = c_0 = 30 \text{ cm}$$

$$b_0 = e = 28 \text{ cm}$$

$$d_0 = \sqrt{a_0^2 + e^2} = 41 \text{ cm}$$

Closed-form solutions exist to relate the final position of the bay tip (x, y, θ) to small changes in the lengths of parts, as shown in Eq. 5.1.

$$\begin{array}{llll} \dfrac{\partial x}{\partial a} = \dfrac{1}{2} & \dfrac{\partial x}{\partial b} = 0 & \dfrac{\partial x}{\partial c} = \dfrac{1}{2} & \dfrac{\partial x}{\partial d} = 0 \\ \dfrac{\partial y}{\partial a} = -\dfrac{a_0}{b_0} & \dfrac{\partial y}{\partial b} = -\dfrac{1}{2} & \dfrac{\partial y}{\partial c} = 0 & \dfrac{\partial y}{\partial d} = \dfrac{d_0}{b_0} \\ \dfrac{\partial \theta}{\partial a} = -\dfrac{1}{b_0} & \dfrac{\partial \theta}{\partial b} = 0 & \dfrac{\partial \theta}{\partial c} = \dfrac{1}{b_0} & \dfrac{\partial \theta}{\partial d} = 0 \end{array} \tag{5.1}$$

Suppose that, in this simplified example, repeatability in the structure is due only to repeatability in the effective length of the members. That is, from deployment to deployment, the struts a, b, c, and d take on different effective lengths, due to changes in seating of the joints. The variability in effective member length from deployment to deployment will be described by its standard deviation $\sigma_a = \sigma_b = \sigma_c = 0.001$ and $\sigma_d = 0.003$, and will be normally distributed in this example. Assuming that the variability of strut length is small in comparison to total strut length, and that there is no correlation between a, b, c, and d, the following approximation holds (this is the error propagation equation for uncorrelated variables [17, p. 41, Eq. 3.14].

$$\sigma_x^2 = \left(\frac{\partial x}{\partial a}\sigma_a\right)^2 + \left(\frac{\partial x}{\partial b}\sigma_b\right)^2 + \left(\frac{\partial x}{\partial c}\sigma_c\right)^2 + \left(\frac{\partial x}{\partial d}\sigma_d\right)^2 \tag{5.2}$$

A sample group of 10 randomly generated bays with members of variable length is shown in Fig. 5.9. For purposes of illustration, the length variability is exaggerated by a factor of 10.

Table 5.1 is an example of how the bookkeeping might be done for this example, and the table shows the final calculated estimates for bay shape repeatability.

Note that finite element analysis is also suitable for relating part variability to shape variability. This type of analysis is not limited to statically determinate structures; as long as a relationship can be built between the measurable variations

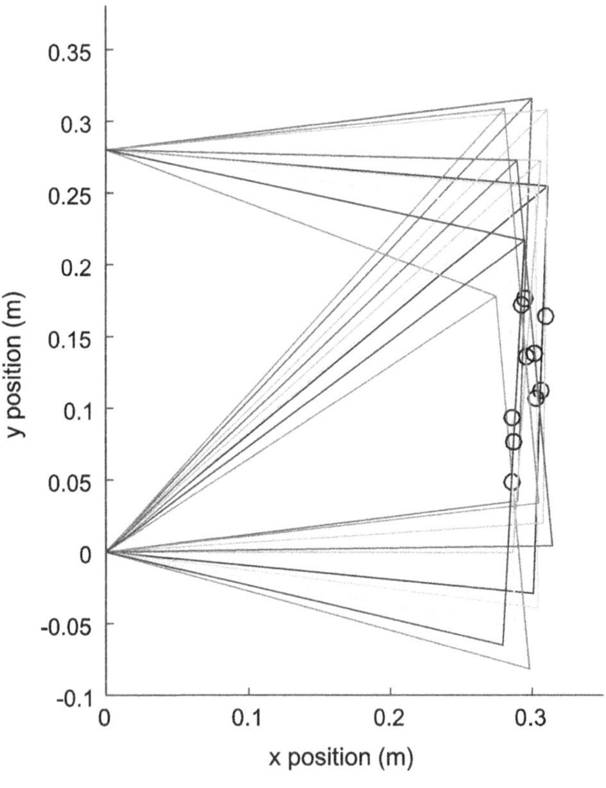

Fig. 5.9 A randomly generated sample of 10 2-D bays (length deviations exaggerated).

TABLE 5.1 EXAMPLE OF A SPREADSHEET DATABASE OF DEPLOYMENT REPEATABILITY FOR THE EXAMPLE BAY

From Design		Measured	From Analysis			Calculated on Spreadsheet		
Element	Nominal Length (m)	Standard Deviation of Length	$\frac{\partial x}{\partial}$	$\frac{\partial y}{\partial}$	$\frac{\partial \theta}{\partial}$	Contribution to RSS: x	Contribution to RSS: y	Contribution to RSS: θ
a	0.30	0.0010	0.50	−1.07	−3.57	2.50E−07	1.15E−06	1.28E−05
b	0.28	0.0010	0.00	−0.50	0.00	0.00	2.50E−07	0.00
c	0.30	0.0010	0.50	0.00	3.57	2.50E−07	0.00E+00	1.28E−05
d	0.41	0.0030	0.00	1.47	0.00	0.00	1.93E−05	0.00
						x	y	θ
			Predicted Repeatability			0.0007	0.0046	0.0051

in the parts and the critical variations of the deployed shape, a parts-based method can be used to analyze deployment repeatability in the structure.

Consider next an assembly of bays into a mast. Figure 5.10 shows a sample group of 10 masts, each formed of 10 randomly generated bays.

When several bays of a truss are to be stacked together in a boom, the relationships between the bay performance and boom performance are also subject to this form of analysis. The boom's shape errors are the deviations x_{boom}, y_{boom}, and θ_{boom} from the nominal tip position. In this example, the variables y and θ from the single bay are correlated, and the total boom tip vertical displacement y_{boom} depends upon both y and θ of each constituent bay. Consequently, calculating the shape repeatability of the full truss is slightly more involved than the equivalent calculations for a single bay. The calculation follows principles of error propagation in correlated variables via the covariance; more detail on the mathematical principles can be found in a statistical methods handbook (such as [17]). Luckily, it is possible to determine the covariance matrices in this example analytically, and there is only one term of interest.

$$x_{\text{boom}} = \sum_{i=1}^{N} x_i \tag{5.3}$$

$$\sigma^2_{x_{\text{boom}}} = N\sigma^2_x \tag{5.4}$$

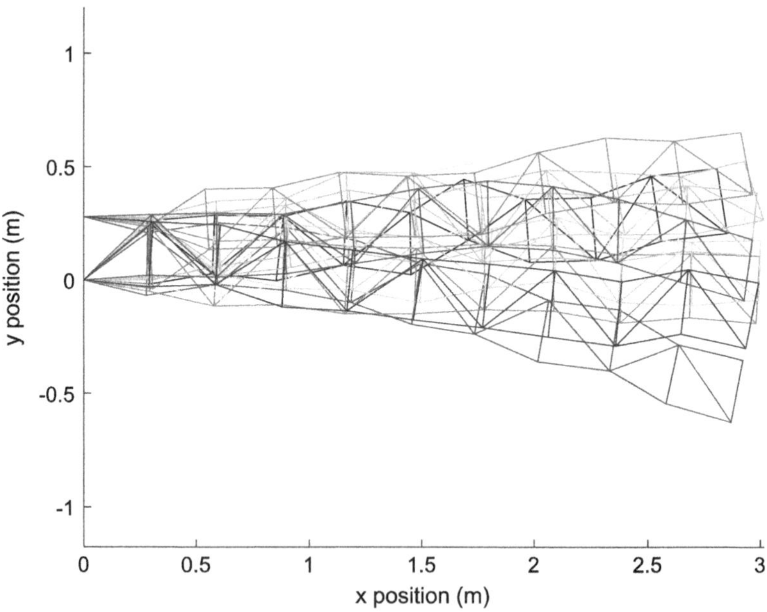

Fig. 5.10 Ten randomly generated 10-bay booms (part length variability exaggerated).

$$\theta_{\text{boom}} = \sum_{i=1}^{N} \theta_i \tag{5.5}$$

$$\sigma^2_{\theta_{\text{boom}}} = N\sigma^2_\theta \tag{5.6}$$

$$y_{\text{boom}} = \sum_{i=1}^{N} y_i + \sum_{i=1}^{N} a_0(N-i)\theta_i \tag{5.7}$$

$$\sigma^2_{y_{\text{boom}}} = N\sigma^2_y + \sum_{i=1}^{N-1}[(N-i)a_0]^2\sigma^2_\theta + 2\sum_{i=1}^{N-1}[(N-i)a_0]\sigma^2_{\theta y} \tag{5.8}$$

$$\sigma^2_{\theta y} = \frac{a_0}{b_0^2}\sigma^2_a \tag{5.9}$$

The results of these calculations are summarized in Table 5.2. For this example structure, a 3 m × 0.28 m truss boom with millimeter-level strut length variability, this simple estimation method predicts cm-level errors in tip lateral position.

If the covariance matrix is not known, it can be approximated from data or modeling. If no paired data exist, or if data are so limited that the covariance is unreliable, consider neglecting the covariance. In the 2-D truss example, assuming (incorrectly) that the covariance between a single bay's errors y and θ is zero introduces an error of 6 percent in the estimation of $\sigma_{y_{\text{boom}}}$. The limited published data on this method suggest a practical error of 100 percent on values like x_{boom}, y_{boom}, and θ_{boom}. It is an order-of-magnitude method, and time spent establishing a highly accurate covariance is likely wasted.

5.4.3 MODELING STATICALLY SUSTAINABLE SHAPES

Consider a structure where repeatability is dominated by joint friction, free play, and other sources of joint hysteresis. The constitutive relationships of the joints, including dead bands and friction, can be experimentally determined, but the assembled structure cannot be deployed in a way that accurately recreates the on-orbit load history in every joint. This suggests a modeling method for bounding the possible shape envelope of a LULS structure: Every joint in a simulation can be individually loaded in the same direction and released, allowing the simulated joint to regain static friction in a new, highly biased position. By evaluating the most extreme statically sustainable deformations of the structure, one can determine a conservative envelope for deployment repeatability.

An example of this sort of analysis can be found in a study by Stohlman and Pellegrino of the ATK (formerly AEC-Able) ADAM mast with a stochastic approach to certain part properties [13, 18]. The work particularly addresses how friction and preload in this structure can affect the range of resting positions of the mast tip when the mast is loaded as a cantilever beam.

TABLE 5.2 EXAMPLE OF A SPREADSHEET DATABASE OF DEPLOYMENT REPEATABILITY FOR THE EXAMPLE BOOM

From Design		Calculated	Contribution to Boom Error				Calculated on Spreadsheet	
Bay Error	Nominal Value	Standard Deviation	x_{boom}	y_{boom}	θ_{boom}	Contribution to RSS: x_{boom}	Contribution to RSS: y_{boom}	Contribution to RSS: θ_{boom}
x	0.0000	0.0007	10.00	0.00	0.00	5.00E−06	0.00E+00	0.00E+00
y	0.0000	0.0046	0.00	10.00	0.00	0.00E+00	2.07E−04	0.00E+00
θ	0.0000	0.0051	0.00	25.65	10.00	0.00E+00	6.54E−04	2.55E−04
		Value						
y, θ covariance		3.827E−06	0.00	27.00	0.00	0.00E+00	1.03E−04	0.00E+00
N		10						
						x_{boom}	y_{boom}	θ_{boom}
				Predicted Repeatability		0.0022	0.0311	0.0160

5.5 CASE STUDIES

Case studies in repeatability cover a wide range of structural types and precisions. This section presents the high-precision repeatability study done for the MOIRE deployable optic and the mid-precision studies of the GEMS and SAILMAST deployable masts. At the end of the section is a collection of other published work on repeatability.

5.5.1 SUB-MILLIRADIAN REPEATABILITY IN MOIRE

Deployable mirror petal supports are among the most-studied structures in the field of repeatability. This is, perhaps, because of the alignment demands placed on a segmented mirror: No source of shape error can be safely neglected. This chapter's summary of case studies, shown in Table 5.3, later in this chapter, includes four such studies, and the MOIRE project is featured here as an example of a particularly demanding application and of repeatability as part of a robust program of characterization.

The MOIRE program is sponsored by the Defense Advanced Research Projects Agency (DARPA) [19]. Figure 5.11 depicts its conceptual approach. The MOIRE system seeks to provide geosynchronous-based Earth observation capabilities. Ball Aerospace & Technologies Corp., with its partners Lawrence Livermore National Laboratory, NeXolve Corporation, and ATK, are designing a 10 m aperture demonstration telescope (ultimately traceable to a future 20 m aperture system) [20, 21] with real-time video downlink capability. The concept would exploit the benefits of transmissive diffractive optics, which allow for approximately three orders of magnitude relaxation in out-of-plane tolerances

Fig. 5.11 Artist's concept of MOIRE based on early efforts in system design [24].

in the primary optic when compared to a reflective system [22]. The primary optic would be stowed for launch and deployed on orbit.

The MOIRE Phase 2 Option 2 study started in August 2012 with the objective of building several hardware models to validate the design concepts. Structural thermal stability results indicate order of magnitude better performance over derived requirements [23], and deployment testing and analysis indicate a robust primary architecture [24].

The MOIRE primary structure consists of 36 diffractive optical elements (DOE) arranged into 18 petal segments that would fold up for launch and then deploy using stored spring energy and eddy current dampers. Figure 5.12

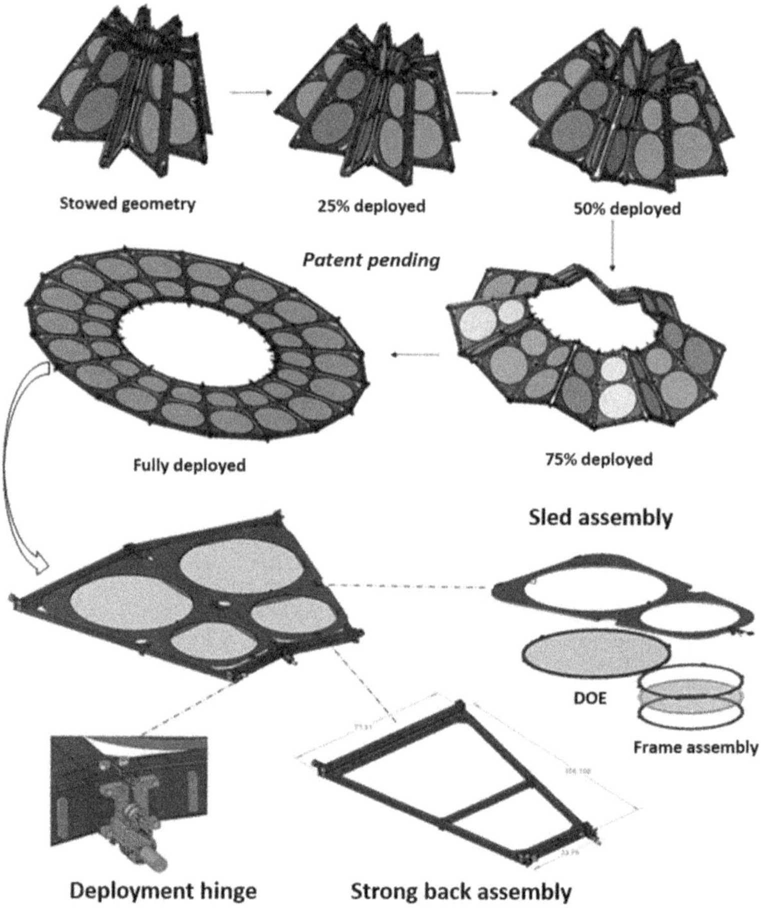

Fig. 5.12 MOIRE primary optical structure design and deployment sequence [24].

shows the deployment sequence and the strong back assembly, sled assembly, and two DOE assemblies within each petal. At full deployment, a semi-kinematic interface provides the features for accurate positioning, illustrated in Fig. 5.13. A FARO Vantage laser tracker system was used to measure target displacements while under representative environmental conditions [25].

A titanium hinge subassembly at each hinge line provides the deployment force required for opening the 18 petals (as shown in Fig. 5.13). The titanium hinge design consists of constant force springs and an eddy current damper. The spring drum design provides tailorable deployment torque via the addition or subtraction of stacked springs while the damper controls the deployment rate. The hinge is mounted via a titanium flex pivot bearing that allows compliance during deployment due to slight hinge misalignments. Final alignment is provided by kinematic features located in the strong back. These features consist of a sphere/cone, sphere/V-groove, and sphere/flat plane to maintain petal-to-petal location and angular placement.

5.5.1.1 MOIRE KINEMATIC MODEL

A Simulink model re-created the DOF and input the physical system parameters such as mass, center of gravity (CG), inertia, torques, damping rates, and friction.

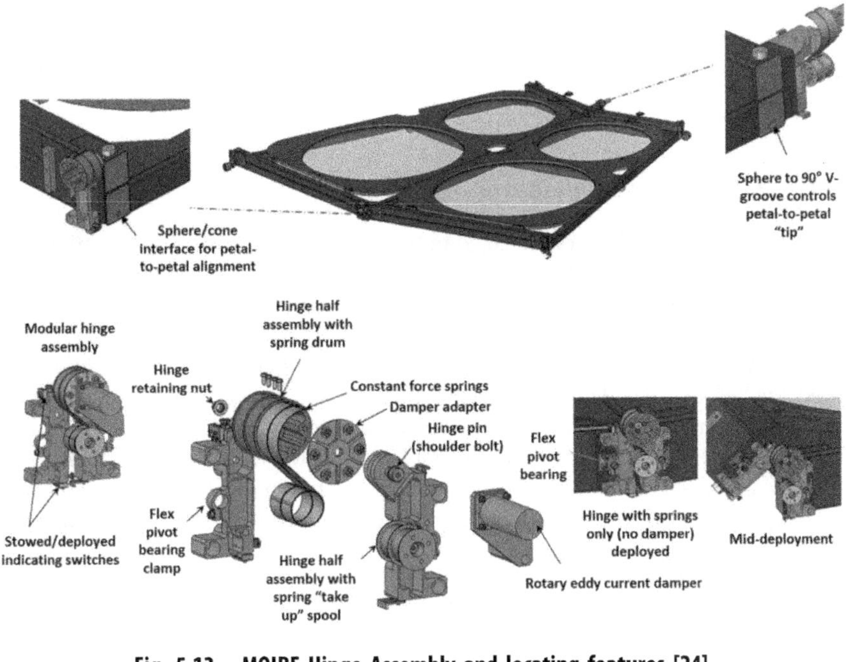

Fig. 5.13 MOIRE Hinge Assembly and locating features [24].

The model served as a virtual test bed to run off-nominal cases, simulating conditions such as varying parasitic torques or failed damper mechanisms. An important result from the Monte Carlo analysis showed there was no reason to introduce additional mechanisms to force synchronous deployment (for example, cables and pulleys) as the system was always driven to full deployment. It also confirmed that no basic geometric constraints exist to prevent full deployment.

A quarter-scale DOE primary model (shown in Fig. 5.14) was constructed to verify the Simulink dynamic analysis results. The scale model made use of commercial off-the-shelf parts and emulated the DOF necessary for deployment. It served as a test bed to investigate effects of different failure modes and to vary system parameters. In addition to verifying robust dynamics and validating the primary optics stowage, the scale model served as a test bed for understanding 0*g* offloader impacts and future design considerations. The design of the offload system can be easily scaled to a 10 m system.

5.5.1.2 FULL-SCALE HINGE DEPLOYMENT MODEL

Two full-scale composite strong back petal segments were fabricated by ATK Magna and joined together with flight-like hinge assemblies. Additional mass simulators and membrane optic simulators were assembled to complete a full-scale deployment model. Mirrors were bonded to adjacent sides of the petal assemblies so that deployment angles could be measured by theodolites. To simulate zero-gravity conditions, the deploying petal used a single air bearing with a spherical joint and adjustable spring load located at the CG, as shown in Figs. 5.15 and 5.16.

5.5.1.3 MOIRE FULL-SCALE DEPLOYMENT TEST RESULTS

The two-petal assembly was subjected to thermal vacuum testing prior to the final deployment measurements, shown in Fig. 5.17. Results indicated the deployment met the requirements with good margin. The average standard deviation from the theodolites was 2.3 μrad, and average deployment time was 145 sec.

Vac-Kote 44147EF + 3% MoS_2 was used at the spherical interfaces and burnished on before each deployment. Examinations of the contact points indicate wear due to ground handling and thermal vacuum testing (shown in Fig. 5.18). An important design feature is adjustable hard stop contact spheres, so fine tuning can be accomplished before launch to compensate for any degradation because the system is a one-time deploy. Further improvements to the design can be made and include

- Create better access to adjustable hard stops.
- Eliminate monolithic fitting and augment the interface with a polished hardened interface, as shown in Fig. 5.18.

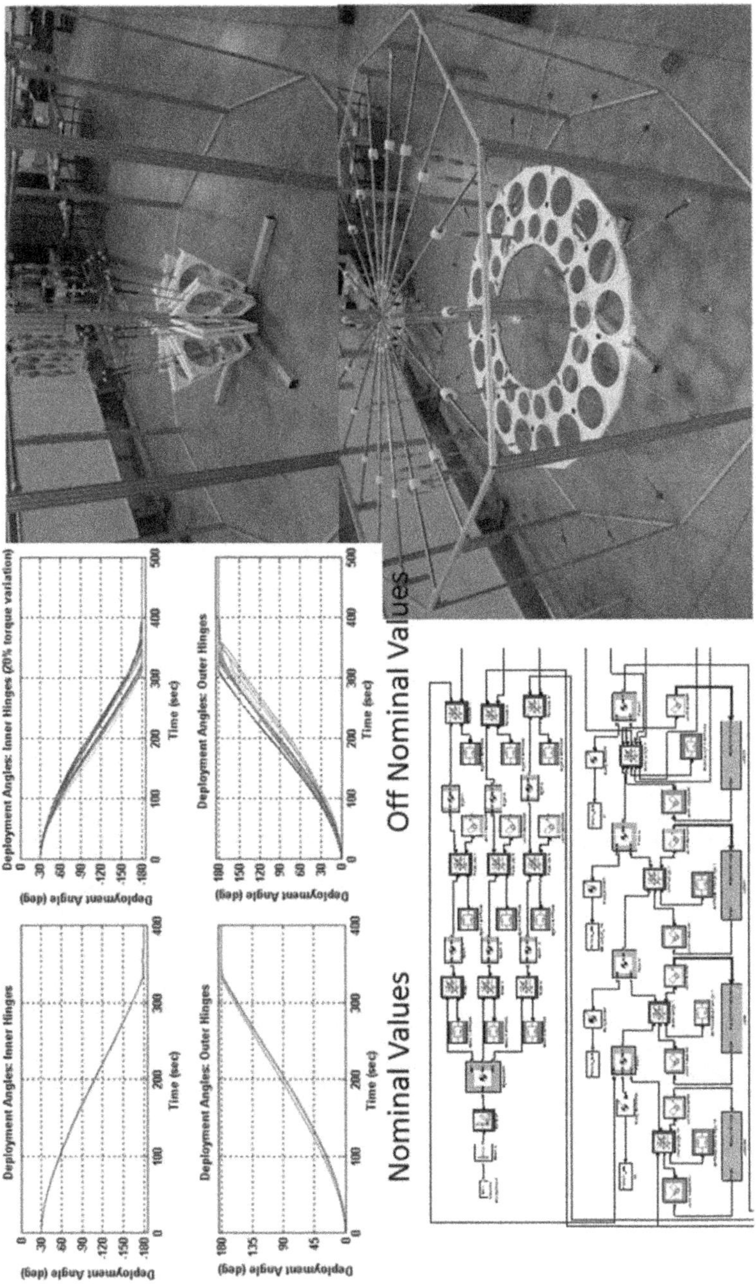

Fig. 5.14 MOIRE Scale Model testing [24].

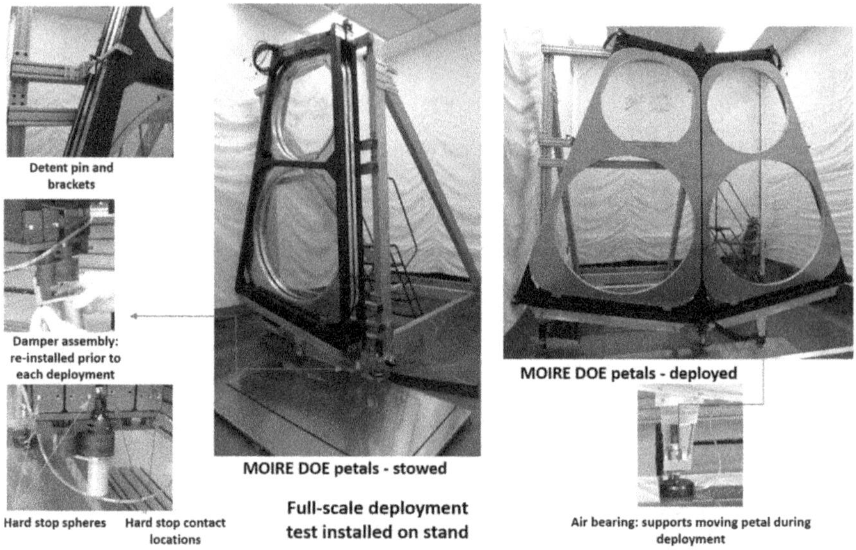

Fig. 5.15 Full-scale deployment testing setup [24].

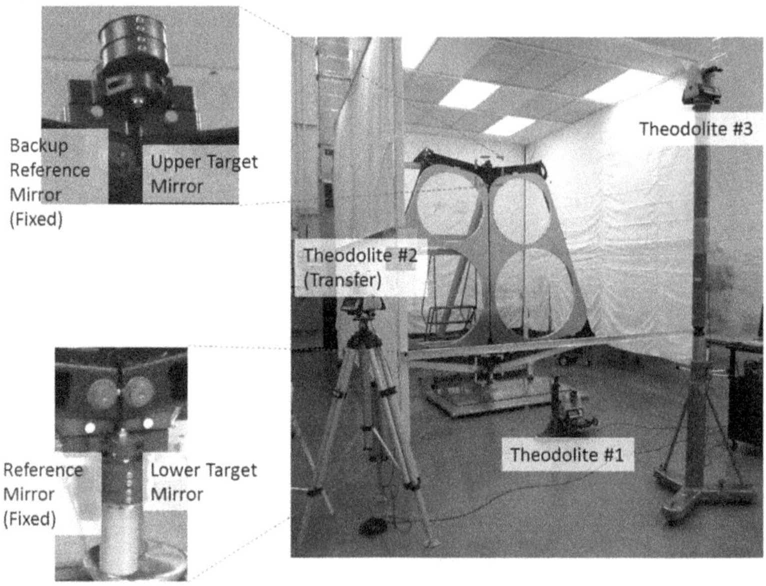

Fig. 5.16 Full-scale deployment testing metrology setup [24].

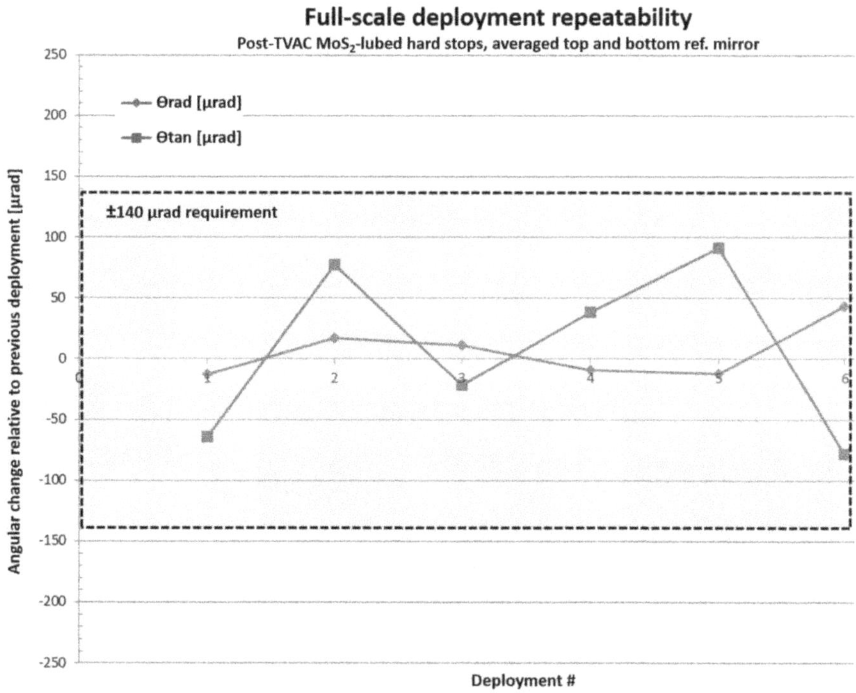

Fig. 5.17 Full-scale deployment repeatability testing [24].

5.5.1.4 MOIRE SUMMARY AND CONCLUSIONS

One of the primary risks in a large space telescope system is the ability to verify total system performance with ground testing. Piece-wise testing must be performed and detailed analytical models to simulate on-orbit performance must be created. MOIRE has demonstrated the capability to meet overall system requirements. The modeling effort showed good correlation; however, subtle details that were not included in the model may have influenced the predictions. These include

- Small manufacturing disparities such as adhesive amount, bond line thickness, fiber volume, and, thus, CTE variability, as shown in coupon testing.
- Thermal strain properties were incorporated; temperature stiffness properties, however, were based on heritage information, not coupon test results.
- Lower-fidelity hinge model.

Finally, it is recommended that next-generation laser tracker equipment be implemented to drive down accuracy errors and to reduce the standard

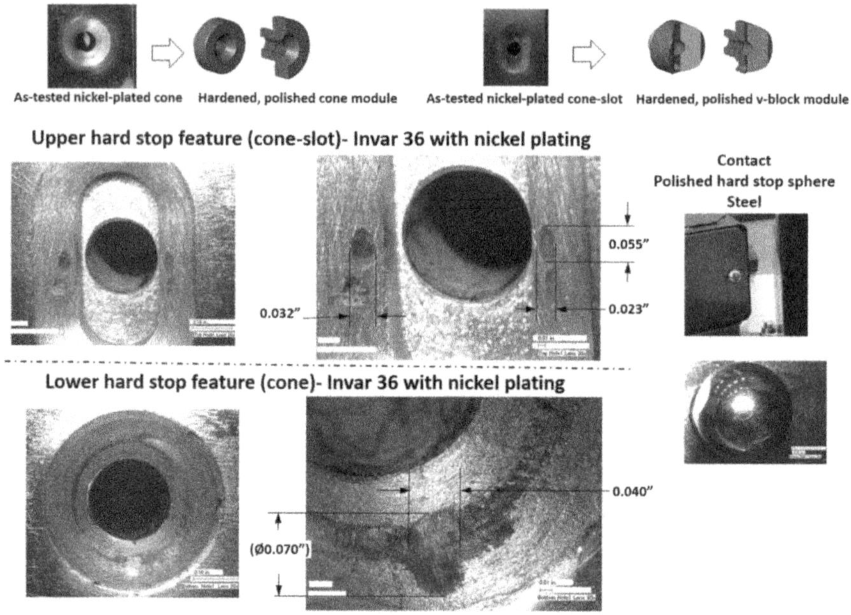

Fig. 5.18 Wear indications at hard stops [24].

deviations of target location. Trackers such as the Nikon Laser Radar do not require specific targets and can directly use structure features, eliminating another error source.

5.5.2 MILLIMETER REPEATABILITY IN SAILMAST AND SUBMILLIMETER REPEATABILITY IN GEMS

McEachen has presented two useful examples of boom testing at ATK in the 40 m SAILMAST project [26] and the 4 m GEMS telescope boom [27]. These studies are straightforward examples of mast test setups that require relatively little specialized equipment.

The GEMS boom had no allowance for on-orbit adjustment and therefore was expected to rely entirely on ground-based demonstration of deployment accuracy and repeatability [27]. It is noteworthy that the GEMS boom was tested for repeatability as a complete structure and evaluated for repeatability via component testing for the statistical repeatability of individual joints [14]. A similar component-based approach is reviewed in Section 5.4.2, earlier in this chapter.

The SAILMAST shape and GEMS tip position were both tracked during ground testing by a laser target tracking system.

Earlier SAILMAST testing in 2005 [28] used photogrammetry to track a set of LEDs in two dimensions, a method that was able to resolve 2.2 mm lateral displacements of a 40 m boom. This method recalls the strategy used by the Shuttle Radar Topography Mission, where a constellation of lights was installed at the tip of the boom and a modified star tracker was used to track tip displacements [29]. This type of target tracking system is especially attractive for structures with very large displacements and millimeter-level precision requirements on the test data.

5.5.3 FURTHER READING

A number of additional studies may be of interest to the reader. Table 5.3 summarizes studies with published repeatability values. It would be unfair to treat this table as an apples-to-apples comparison of performance, as few of these structures have the same figures of merit, and not all report the same statistical measure of repeatability. Deployment repeatability experiments are rare, so many structures are characterized by their hysteretic limits or system requirements.

Much has been published on the Shuttle Radar Topography Mission (SRTM) mast. Data on its repeatability is found in Gross and Messner's paper on the technology of the ADAM mast [30]; this is a high-level presentation on the mast's performance without a detailed presentation of test methods. The metrology system, which is also an interesting subject in shape accuracy, is described well in a separate publication [29], and both the SRTM and somewhat different NuSTAR [3] metrology systems may be of interest for studies of mast tip positioning.

Frank and Dumm [31] used a granite-table-mounted test setup to evaluate the deployment repeatability of a large tripod hinge with strain-energy-deployed tape-spring hinges. The tip of the deployable truss, designed to support an optical mirror, was instrumented with a set of three orthogonal lasers. These lasers projected points into two-dimensional photodiode arrays, allowing six-degree-of-freedom tracking of the tip position. Deployment repeatability was found to be in the hundreds of microns over 2.9 m. The introductory section of this paper is an excellent overview of useful case studies on elastic composite hinge repeatability and analysis. A similar metrology concept was used by Black et al. in evaluating the repeatability of composite tapespring hinges and a deployable tripod [32]. Both papers have clear mathematical descriptions of how tip position was derived from laser dot position.

Hysteresis in composites is a subject outside the scope of this book. Temperature-dependent relaxation and creep, plastic deformation, and progressive damage may each be significant and may be amenable to analysis. With good-quality material characterization and environment prediction, the uncertainty lies in the storage time and launch delays. Domber et al. [10] have studied a folding composite hinge for repeatability, demonstrating an initial break-in effect during the first deployment of a composite hinge, and then apparently random behavior and repeatable creep recovery thereafter. The design of the hinge is proprietary, but the study reviews many of the issues of stowage time, creep, break-in, and temperature dependence that are likely to arise in the study of any composite

TABLE 5.3 EXAMPLES OF DEPLOYABLE STRUCTURES WITH CHARACTERIZED REPEATABILITY

Case Study Examples and Literature References	Structural Class—Level of Assembly	Structural Class—Critical Dimension	Structural Class—Dimensional Precision	References	Characteristic Repeatability (note that certain sources report the 3-sigma range and others report the standard deviation)	Special Equipment
SRTM mast	Assembly	1-D	Moderate	[29, 30]	Length: 60 m $\pm$ 1.3 mm Lateral: $\pm$13 mm Twist: $\pm 0.13^\circ$ Bending: $\pm 0.005^\circ$	
GEMS mast	Assembly	1-D	Moderate	Section 5.5.2 [14, 27]	Length: 0.85 m Lateral: $\pm$0.096 mm Torsional: $\pm$2.1 arcsec	Spreadsheet estimation
WMAP cable latches	Component	1-D	Moderate	[13, 18]	Cable stroke: 140 $\pm$ 0.01 mm	Material testing machine
IHOSS hinge truss [31]	Component	1-D	Moderate	[31]	Angle range: 60° Tip position: $<\pm$0.5 mm Tip angle: $<\pm 0.11^\circ$	Granite table, 2-D photodiodes
Composite hinges [32]	Component	1-D, 3-D	Moderate	[32]	Angle range: 180° Lateral repeatability: $\pm$50 μm	
Composite hinges [10]	Component	1-D	Moderate	[10]	Angle ranges: 150–164°[1] Axial: $\pm$0.0–2.3 μm Lateral: $\pm$0.1–9.0 μm	

Composite hinges [33]	Component	1-D	Moderate	[33]	Angle range: 135° Axial: $< \pm 1$ μm Roll: $< \pm 8$ μrad Pitch: $< \pm 8$ μrad	Thermal chamber, Invar reference
MADE joint	Component	1-D	Moderate	[34]	Angle range: 180° Lateral: $< \pm 1$ μm	
MOIRE petals	Subassembly	3-D	Extreme	Section 5.3.3 [23, 35, 36, 37]	Angle: $\pi \pm 2.3$ μrad	Theodolite
Ball bearing hinge-deployed mirror	Assembly	3-D	Moderate	[38]	Angle range: 40° Hinge axis: ± 13 μm Deployment direction: ± 35 μm	Eddy current probes (ECPs)
ESDM petal support	Subassembly	3-D	Extreme	[5]	Latch angle range: 93° Tip position: ± 1.2 μm	
Lidar prototype petal support	Subassembly	3-D	Extreme	[15]	Latch angle range: 51° Hinge axis: ± 0.81 μm Deployment direction: ± 8.6 μm	
Lockheed test petal	Subassembly	3-D	Extreme	[39]	Angle range: 90° Tip position: <71 μm	

hinge. Silver et al. [33] have also presented a meticulous test setup for high-precision data on composite hinges in thermal testing. Hysteresis and randomness are both carried as finite element analysis component properties in Tanaka and Saito's study of a simple cable network [40].

A 2002 paper by Heald and Peterson [15] opens with a history of related joint repeatability studies, a series that focuses on the one-dimensional deployment and latching of a large mirror or mirror segment at a hinge. Heald and Peterson compare a lidar segment deployer with Warren et al.'s earlier Engineering Science Development Model (ESDM) telescope petal deployer [5], a study that includes data on post-deployment stability and microlurching. The Micron level Accuracy Deployment Experiment (MADE) was a proposed [34] in-space technology demonstration. It was intended to demonstrate precision deployment for segmented optics and other high-precision LULS structures, and to reduce the perceived risk of deployment mechanisms in space. The ball bearing hinge included in the MADE proposal and noted in Lake et al.'s 1998 paper on deployment repeatability [1] was also evaluated alongside a lower-performance bearing hinge by Kozola et al. [38] This competitive evaluation of the hinges goes into some detail about accommodation of the thermal environment and realistic loading conditions.

Inflatables are not well studied for shape repeatability, but a 2000 paper by Pappa et al. [41] does describe the repeatability and precision of a photogrammetry system in the characterization of a 5 m diameter inflatable antenna.

5.6 SUMMARY

Simply by being large, LULS magnify sources of shape error that are negligible in smaller spacecraft. Every deployable element adds some additional variability to the deployed shape. The design elements that can compensate for shape errors, like alignment mechanisms and rigid parts that support high preloads, are traded against stringent mass requirements. At the same time, applications like telescope alignment and even solar sail thrust generation have inherent shape-quality requirements. For these reasons, deployment repeatability is a higher risk for LULS than for spacecraft with less complex deployables.

At the same time, large ultra-lightweight structure deployment puts remarkable demands on test equipment. The end result is that repeatability testing of these structures must use some combination of particularly well-characterized test equipment and judicious modeling.

When one is approaching repeatability in a project, it is best to be meticulous about bookkeeping. Issues of modeling uncertainty and post-deployment stability contribute to the same error budget as repeatability concerns, so be clear about which shape error sources will be considered under each heading to avoid neglecting or double-counting these effects. Within a given structure, one of these effects may be so dominant that the others are unimportant, and a clear accounting of shape error contributors can save project effort in the long run.

When one is evaluating repeatability for comparison with a shape error requirement, preference should be given to testing over modeling and to the largest possible assembly that can be tested. The stability of test fixtures and sensors should exceed the repeatability of the test article.

As a final case study in managing deployment repeatability, the NuSTAR x-ray telescope used a 10-m-long deployable mast to separate its two arrays of focusing mirrors from its two detectors [3]. Early in the development of the system, questions about both the accuracy of the deployed shape during gravity-compensated tests and deployment repeatability were raised. At this point, there were two possible paths: testing to strictly bound the repeatability, combined with both testing and modeling to increase confidence in shape accuracy; or a metrology and adjustment system to correct for any errors.

The NuSTAR project opted for a metrology and adjustment system, which had many advantages for the project and few disadvantages. Because a three-degree-of-freedom alignment between the root and tip of the mast was the only adjustment required, a laser metrology system and angle adjustment mechanism were sufficient. The mast deployment was ultimately tested with alignment monitoring only four times before flight, and alignment requirements were met during commissioning on orbit. Simpler testing for mechanical deployment, without alignment monitoring, was also performed. In addition to greatly reducing the requirements for the mast's accuracy and repeatability, the alignment mechanism was used to remove misalignments from thermal distortion of the mast throughout the course of the mission [42].

Where a post-deployment adjustment is feasible, it can greatly reduce the need for deployment testing. The disadvantages of such a system are typically added complexity, mass, volume, and commissioning time. Deployment repeatability can be traded against many other elements of spacecraft performance, and the cost and schedule burden of repeatability testing can be included in that trade with good planning.

ACKNOWLEDGMENTS

The authors are grateful to the editors and to Michael McEachen (Orbital ATK), Yunjin Kim (JPL), and Bill Craig (UC Berkeley) for their consultation on this chapter.

REFERENCES

[1] Lake, M. S., Peterson, L. D., Hachkowski, M. R., Hinkle, J. D., and Hardaway, L. R., "Research on the Problem of High-Precision Deployment for Large-Aperture Space-Based Science Instruments," in *Space Technology & Applications International Forum*, Albuquerque, NM, 1998.

[2] Figliola, R. S., and Beasley, D. E., *Theory and Design for Mechanical Measurements, Third Edition*, John Wiley & Sons, NY, 2000.

[3] Liebe, C. C., Burnham, J., Cook, R., Craig, B., Decker, T., Harp, D. I., Kecman, B., Meras, P., Raffanti, M., Scholz, C., Smith, C., Waldman, J., and Wu, J., "Metrology System for Measuring Mast Motions on the NuSTAR Mission," *IEEE Aerospace Conference*, Big Sky, MT, 2010.

[4] Pappa, R. S., Black, J. T., Blandino, J. R., Jones, T. W., Danehy, P. M., and Dorrington, A. A., "Dot-Projection Photogrammetry and Videogrammetry of Gossamer Space Structures," *Journal of Spacecraft and Rockets*, Vol. 40, No. 6, 2003, pp. 858–867.

[5] Warren, P. A., Peterson, L. D., and Hinkle, J. D., "Submicron Mechanical Stability of a Prototype Deployable Space Telescope Support Structure," *Journal of Spacecraft and Rockets*, Vol. 36, No. 5, 1999, pp. 765–771.

[6] Ingham, M. D., *Microdynamics and Thermal Snap Response of Deployable Space Structures*, MIT, Cambridge, MA, 1998.

[7] Silver, M. J., Peterson, L. D., and Hardaway, L. R., "Experimental Measurement of Picometer Scale Spontaneous Vibrations in a Precision Deployable Boom under Thermal Loading," *Shock and Vibration*, Vol. 14, No. 2, 2007, pp. 133–149.

[8] Domber, J. L., Atcheson, P., and Kommers, J., "MOIRE: Ground Test Bed Results for a Large Membrane Telescope," *AIAA SciTech*, National Harbor, MD, 2014.

[9] Hardaway, L. R., and Peterson, L. D., "Ground-based Microdynamic Testing of Deployed Optical Structures," *AIAA Structures, Structural Dynamics and Materials Conference*, St. Louis, MO, 1999.

[10] Domber, J. L., Hinkle, J. D., Peterson, L. D., and Warren, P. A., "Dimensional Repeatability of an Elastically Folded Composite Hinge for Deployed Spacecraft Optics," *Journal of Spacecraft and Rockets*, Vol. 39, No. 5, 2002, pp. 646–652.

[11] Lake, M. S., Warren, P. A., and Peterson, L. D., "A Revolute Joint with Linear Load-Displacement Response for Precision Deployable Structures," *AIAA Structures, Structural Dynamics, and Materials Conference*, Salt Lake City, UT, 1996.

[12] Atkinson, C., Gilman, L., and Reynolds, P. "Technology Development for Cryogenic Deployable Telescope Structures and Mechanisms," *Proceedings of SPIE, Vol. 5179, Optical Materials and Structures Technologies*, San Diego, CA, 2003.

[13] Stohlman, O. R., and Pellegrino, S., "Effects of Component Properties on the Accuracy of a Joint-dominated Deployable Mast," *AIAA Structures, Structural Dynamics, and Materials Conference*, Denver, CO, 2011.

[14] McEachen, M. E., "Development of the GEMS Telescope Optical Boom," *AIAA Structures, Structural Dynamics, and Materials Conference*, Denver, CO, 2011.

[15] Heald, J. C., and Peterson, L. D., "Deployment Repeatability of a Space Telescope Reflector Petal," *Journal of Spacecraft and Rockets*, Vol. 39, No. 5, 2002, pp. 771–779.

[16] Heald, J. C., and Peterson, L. D., "Sensitivity of the Deployment Repeatability of a Precision Telescope Mechanism," *AIAA Structures, Structural Dynamics, and Materials Conference*, Denver, CO, 2002.

[17] Bevington, P. R., and Robinson, D. K., *Data Reduction and Error Analysis for the Physical Sciences*, McGraw-Hill, NY, 2002.

[18] Stohlman, O. R., *Repeatability of Joint-Dominated Deployable Masts*, PhD Thesis, California Institute of Technology, Pasadena, CA, 2011.

[19] *MOIRE - Membrane Optic Imager Real-Time Exploration*, Solicitation number: DARPA-BAA-10-51, 2010.

[20] Hyde, R. A., "Eyeglass. 1. Very Large Aperture Diffractive Telescopes," *Applied Optics*, Vol. 38, No. 19, 1999, pp. 4198–4212.

[21] Early, J. T., Hyde, R., and Baron, R. L., "Twenty-Meter Space Telescope Based on Diffractive Fresnel Lens," *Proc. SPIE*, Vol. 5166, San Diego, CA, 2004.

[22] Atcheson, P. D., and Kommers, J., "MOIRE: Ground Test Bed Results for a Large Membrane Telescope," *Scitech*, National Harbor, MD, 2014.

[23] Domber, J. L., Belnap, B., Rynders, R., and Schweickart, R., "MOIRE Strongback Thermal Stability Analysis and Test Results," *AIAA Scitech*, Kissimmee, FL, 2015.

[24] Waller, D., Campbell, L., Domber, J. L., Putnam, D., and Thompson, R., "MOIRE Primary Diffractive Optical Element Structure Deployment Testing," *AIAA Scitech*, Kissimmee, FL, 2015.

[25] Pfeffer, C., "FARO Technical White Paper Topic: ADM (Absolute Distance Measuring)," FARO Technologies, 2006.

[26] McEachen, M. E., "Validation of Slender Lattice Trusses by Modeling and Testing," *Journal of Spacecraft and Rockets*, Vol. 49, No. 6, 2012, pp. 970–977.

[27] McEachen, M. E., "Verification of Placement Precision and Stability for the GEMS Telescope Optical Boom," *54th Structures, Structural Dynamics, and Materials Conference*, Boston, MA, April 8–11, 2013.

[28] McEachen, M. E., Trautt, T. A., and Murphy, D. M., "The ST8 SAILMAST Validation Experiment," *46th AIAA Structures, Structural Dynamics, and Materials Conference*, Austin, Texas, 2005.

[29] Duren, R., Wong, E., Breckenridge, B., Shaffer, S., Duncan, C., Tubbs, E., and Salomon, P., "Metrology, Attitude, and Orbit Determination for Spacebourne Interferometric Synthetic Aperture Radar," *SPIE AeroSense Conference on Acquisition, Tracking, and Pointing XII*, 1998.

[30] Gross, D., and Messner, D., "The Able Deployable Articulated Mast-Enabling Technology for the Shuttle Radar Topography Mission," *33rd Mechanisms Symposium*, Pasadena, CA, 1999.

[31] Frank, N. W., and Dumm, H. P., "Deployment Repeatability of a Truss Structure Utilizing Integral Composite Hinges," *AIAA Structures, Structural Dynamics, and Materials Conference*, Newport, RI, 2005.

[32] Black, J. T., Whetzal, J. A., deBlonk, B. J., and Massarello, J. J., "Deployment Repeatability Testing of Composite Tape Springs For Space Optics Applications: *47th AIAA Structures, Structural Dyanamics, and Materials Conference*, Newport, RI, 2006.

[33] Silver, M. J., Echter, M. A., Reid, B. M., and Banik, J. A., "Precision High-Strain Composite Hinges for the Deployable In-Space Coherent Imaging Telescope," *AIAA Scitech*, San Diego, CA, 2016.

[34] Peterson, L. D., and Lake, M. S., "Micron Accuracy Deployment Experiment (MADE)," NASA technical report NASA-CR-197792, 1995.

[35] Atcheson, P., Domber, J., Whiteaker, K., Britten, J. A., Dixit, S. N., and Farmer, B., "MOIRE: Ground Demonstration of a Large Aperture Diffractive Transmissive Telescope," *Proc. SPIE 9143, Space Telescopes and Instrumentation*, Montreal, Canada, 2014.

[36] Atcheson, P. D., Stewart, C., Domber, J., Whiteaker, K., Cole, J., Spuhler, P., Seltzer, A., Britten, J. A., Dixit, S. N., Farmer, B., and Smith, L., "MOIRE: Initial Demonstration of a Transmissive Diffractive Membrane Optic for Large Lightweight Optical Telescopes," *Proc. SPIE*, Vol. 8442, Amsterdam, Netherlands, 2012.

[37] Waller, D., Domber, J., Price, C., Schweickart, R., Thompson, R., and Whiteaker, K., "MOIRE Thermal Vacuum Structural Stability Testing," *AIAA Scitech*, Kissimmee, FL, 2015.

[38] Kozola, B. D., Erwin, R. S., Robertson, L. M., Griffin, S. F., and Powers, M. K., "Experimental Characterization of Deployment Repeatability for Optical-Precision Space Structures," *42nd AIAA Structures, Structural Dynamics, and Materials Conference*, Seattle, WA, 2001.

[39] Carrier, A. C., Aubrun, J. N., Clappier, R. R., Hilby, T., Lorell, K. R., Romney, B., Sokolsky, L., and Uphoff, J., "Development and Microdynamics Characterization of a Deployable Petal Assembly At Full Scale," *Proc. SPIE 4950, IR Space Telescope and Instruments 254*, 2003.

[40] Tanaka, H., and Saito, K., "A Study of the Shape Repeatability of a Cable-Network Structure," *28th International Symposium of Space Technology and Science*, Ginowan City, Okinawa, Japan, 2012.

[41] Pappa, R. S., Giersch, L. R., and Quagliaroli, J. M., "Photogrammetry of a 5 m Inflatable Space Antenna with Consumer Digital Cameras," NASA Technical Memorandum TM-2000-210627, Washington, DC, 2000.

[42] Interview with Kim, Y., Feb. 10 2016.

CHAPTER 6

Structural Dynamics

Jonathan T. Black*

Virginia Polytechnic Institute and State University, Blacksburg, VA

NOMENCLATURE

$\mathbf{M}$	system mass matrix
$\mathbf{C}$	system damping matrix
$\mathbf{K}$	system stiffness matrix
$\mathbf{x}$	system position vector
$\dot{\mathbf{x}}$	system velocity vector
$\ddot{\mathbf{x}}$	system acceleration vector
$\mathbf{f}(t)$	external force applied to the system
$H(\omega)$	frequency response function
$X(\omega)$	Fourier transform of the input
$Y(\omega)$	Fourier transform of the output
ω_i	ith system structural natural frequency

6.1 INTRODUCTION

Modal analysis is defined as the study of the dynamic characteristics of a mechanical structure. These characteristics are studied through modal testing, which involves collecting and analyzing dynamic data. Mathematically, the process involves relating the input or excitation to a structure to the output or measured response. The theory assumes that any response that differs from the excitation is attributable to the structural dynamics of the structure under test, revealing its natural frequencies, mode shapes, and structural damping. Modal testing and analysis techniques were developed originally for stiff, massive structures to which excitation may be directly applied and responses measured without altering or affecting the structure itself. This chapter describes experimental modal testing and analysis techniques, also called *structural dynamics testing*, specifically adapted to the challenges of ultra-lightweight spacecraft. Table 6.1 provides an overview of the case studies discussed here.

---

*Professor, AIAA Associate Fellow

TABLE 6.1 MODAL TESTING OF ULTRA-LIGHTWEIGHT SPACECRAFT STRUCTURES AND COMPONENTS

Test Goal	Excitation Method	Measurement Method	Appropriate Structural Class—Level of Assembly	Appropriate Structural Class—Critical Dimension	Appropriate Structural Class—Dimensional Precision	Test Article Examples
Freq., mode shape, and damping	Contact	Non-contact	Assembly	2D	Least	SAFE (6.4.1.1)
Freq., mode shape, and damping	Contact	Non-contact	Assembly	2D	Least	SAFDE (6.4.1.1)
Freq., mode shape, and damping	Contact	Contact	Component	1D	Moderate	Tubes (6.4.1.2)
Freq. and mode shape	Contact	Non-contact	Assembly	2D	Least	NGST Sunshield (6.5.1)
Freq., mode shape, and damping	Contact	Non-contact	Subassembly	2D	Least	2 m Sail Quadrants (6.5.1)

Freq., mode shape, and damping	Contact	Contact	Component	2D	Moderate	MFC (6.5.1)
Freq. and mode shape	Contact	Non-contact	Subassembly	3D	Moderate	5 m Inflatable Reflector (6.5.1)
Freq. and mode shape	Contact	Non-contact	Subassembly	2D	Moderate	Pressurized tori (6.5.2)
Freq., mode shape, and damping	Contact and non-contact	Contact and non-contact	Component	2D	Moderate	Thermal-formed membrane panels (6.5.2)
Freq., mode shape, and damping	Contact	Non-contact	Assembly	3D	Extreme	Hexapod (6.5.2)
Freq. and mode shape	Contact	Non-contact	Subassembly	2D	Extreme	Optical Membranes (6.5.2)

6.2 FUNDAMENTALS OF MODAL TESTING

To understand how and why the dynamic tests discussed here were successful, a brief summary of the theory is presented. This summary is based on Refs. [1–11].

The dynamic behavior of any structure is described by its equation of motion, which take the form of Eq. (6.1) for multi degree of freedom (MDOF) systems:

$$\mathbf{M}\ddot{\mathbf{x}} + \mathbf{C}\dot{\mathbf{x}} + \mathbf{K}\mathbf{x} = \mathbf{f}(t) \tag{6.1}$$

Equation (6.1) can be solved by the eigensolution of the following reorganized equation assuming negligible damping and harmonic motion of the form $x_1 = A_1\sin(\omega t) = A_1 e^{i\omega t}$:

$$[\mathbf{K} - \omega^2\mathbf{M}]\mathbf{x} = \mathbf{f}(t) \tag{6.2}$$

where the eigenvalues (roots) of $[\mathbf{K} - \omega^2\mathbf{M}]$ are the undamped structural natural frequencies of the system and the eigenvectors are displacement vectors called *mode shapes* unique to each frequency. The structural natural frequencies and mode shapes describe the dynamic behavior of any system, and their extraction is the objective of the experimental modal analysis process.

Traditionally, structural natural frequencies and mode shapes are experimentally identified through the calculation of frequency response functions. A frequency response function (FRF) $H(\omega)$ can be calculated by relating a known excitation to a known response of the system under test. Peaks in the frequency domain and corresponding phase shifts of the FRF indicate damped structural natural frequencies of the system. The FRF can be generalized as the Fourier transform of the output divided by the Fourier transform of the input, or $H(\omega) = Y(\omega)/X(\omega)$, although it may be computed in several ways depending on the objective, such as to minimize noise, among others [9].

Once FRFs have been collected experimentally, they are loaded into a modal analysis software package that fits the data in a process called *modal parameter estimation* (MPE). Several iterative algorithms were developed in the late 1970s and early 1980s to perform MPE on large MDOF systems [10–12]. The result of the fit is the transfer function of the system, the poles of which correspond to structural natural frequencies. The system response at each pole can then be animated, showing the mode shape that corresponds to the structural natural frequency.

6.3 ADAPTING STRUCTURAL DYNAMICS TESTING TECHNIQUES TO LARGE ULTRA-LIGHTWEIGHT SPACECRAFT (LULS)

The modal test methods of choice for many LULS structures are videogrammetry or laser vibrometry. *Videogrammetry* is the process of applying photogrammetry to synchronized video sequences. The photogrammetry processing is applied at each video frame, generating time histories of a set of three-dimensional points. Videogrammetry is a non-contact, full-field measurement technique. *Laser vibrometry* is also inherently non-contact but analyzes single points on the structure instead of the structure as a whole. The resolution of a videogrammetry measurement is identical to that of a photogrammetry measurement, but because digital video cameras are generally lower resolution than their static counterparts, the measurements also tend to have lower resolution. In general, most videogrammetry measurements will have a resolution of 1:1000, or 1/1000 the size of the object being imaged. The resolution of laser vibrometers is not dependent on the structure but is a property of the laser. These systems generally have a resolution of 10 μm. Because the result of a videogrammetry measurement is a three-dimensional time history of a set of points, the technique is inherently three dimensional, meaning only one set of data is required to analyze in- and out-of-plane motion. Conversely, a laser vibrometer can measure only one degree of freedom (DOF), along the laser beam.

When one is measuring the lightest and largest LULS structures, solar sails, and sunshields comprised of ultra-thin membranes potentially thousands of square meters in size, attaching objects directly to the membranes (including small retro-reflective targets) substantially alters the local mass and stiffness of the test article and may invalidate the data. Therefore, a technique of projecting targets onto the membranes was developed [13–16] to enable the use of videogrammetry or laser vibrometry to characterize the dynamic behavior of such LULS membrane structures.

Despite the mass savings, several factors complicate the use of dot-projection photogrammetry and videogrammetry. The first involves the projection of a target field over the thousands of square meters of surface area of the membranes. Two notable solutions to this problem were developed. One involved doping the membrane material with a dye that fluoresces when exposed to a particular wavelength of incident laser light. A laser could project any type of grid on any part of the membrane and could change the grid or grids over the course of the mission. The other solution was to use the reflection of stars off the specular surface of the membrane as targets [14, 17].

The second complication with using dot-projection videogrammetry is that the targets are not attached to the surface, meaning the measurement technique is no longer three dimensional. Only motion along the ray of light can be measured, identical to a laser vibrometry measurement [16].

A third complication related to the use of videogrammetry in general, not only dot-projection videogrammetry, is that only the response of the structure is measured. To create frequency response functions that allow complete characterization of the structural natural frequencies, mode shapes, and damping of the test article, simultaneous measurement of the excitation supplied to the system and the system response is necessary. While it is possible to independently measure the excitation and to create separate FRFs, it may be difficult to reliably synchronize two different metrology systems. Conversely, the single data acquisition system used in laser vibrometer, impact hammer, and other more standard dynamic testing automatically synchronizes the input and response measurements and creates FRFs in a single step. The temporal videogrammetry data can be filtered and animated to view the motion of the structure at particular frequencies, called operating deflection shapes (ODS), but these cannot be considered fundamental dynamic properties of the system.

Questions about the relationship between ODS, their corresponding frequencies, and structural natural frequencies are common for highly-damped structures such as these. References [13, 15, 18] directly compared structural natural frequencies and mode shapes extracted from FRFs measured with laser vibrometry to operating deflection shapes and the corresponding frequencies extracted from temporal filters and Fourier transforms of the time histories measured with dot-projection videogrammetry. The test article was one triangular quadrant of a small 4 m^2 four-quadrant solar sail test article. To facilitate the dot-projection technique, diffuse white 0.1-mm thick drafting Vellum was used in place of actual aluminum-coated Kapton® or Mylar membranes. The projected target field that was recorded by the video cameras was also used as the measurement grid of the scanning laser vibrometer. Excitation was applied by a shaker attached to the tip of one of the two booms from which the quadrant was suspended. The right-triangular Vellum quadrant was suspended at three attachment points at the corners of the triangle. The right-angle corner was attached to the sail hub, and the two other corners were attached to the tips of hollow steel rods (OD 0.47 cm, ID 0.30 cm) or booms extending from the hub.

The FRFs generated by the laser vibrometer showed five structural natural frequencies occurring between 1 and 5 Hz, as did the operating deflection dot-projection videogrammetry measurement. The discrepancy in the measurements, with the structural natural frequencies taken as actual, was calculated to be 5.14 percent, 1.15 percent, 3.89 percent, 1.36 percent, and 0.021 percent, respectively. This comparison indicates that although true FRFs were not measured, the frequencies corresponding to the operating deflection shapes obtained using dot-projection videogrammetry measurements are extremely close to the actual structural natural frequencies of the structure. The visual comparison of the mode shapes to the operating deflection shapes showed them to be identical, as seen in fifth mode shapes shown in Fig. 6.1. Therefore, the authors concluded that although no FRFs are calculated, dot-projection videogrammetry gives a very accurate picture of the dynamic behavior of membrane-based LULS structures, and the technique could be very useful in future gossamer structure investigations or missions.

A key challenge of testing LULS is their inability to self-support in 1g. Gravity presents both structural and deployment testing challenges. Several important remediations to the challenges presented by gravity are discussed in Chapter 4.

6.4 HISTORICAL MODAL TESTING OF LARGE ULTRA-LIGHTWEIGHT (GOSSAMER) SPACECRAFT STRUCTURES

The development of the Eigensystem Realization Algorithm (ERA) and other iterative time-domain modal parameter estimation techniques in the mid- to

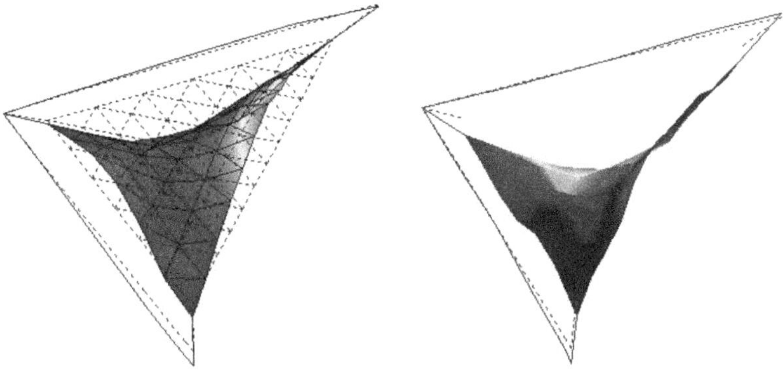

Fig. 6.1 Fifth mode of Vellum solar sail quadrant [15, 18]. (a) Laser vibrometry shape at 4.74 Hz. (b) Videogrammetry shape at 4.75 Hz.

late-1970s is the foundation upon which modal testing of large space structures, including ultra-lightweight space structures and automotive systems, was built [12]. These iterative techniques, and the advances in computer processor speeds that made them feasible, enabled accurate and rapid estimation of the structural natural frequencies, mode shapes, and damping of large, complex, and nonlinear MDOF structures. Hedgepeth [19] described critical requirements for making large flexible space structures feasible, including dynamic testing and data analysis methodologies. Juang and Pappa [12] state that "Active control of flexible structures will require the combined efforts of researchers in both [modal testing and system identification for controls] disciplines," and they present an exhaustive survey of the history of modal testing from the 1940s through the late 1980s. Reference [20] presents a survey of experimental vibration testing of membrane structures.

6.4.1 LEAST DIMENSIONAL PRECISION

INTRODUCTION

LULS that fall into the category of those requiring the least dimensional precision, including solar sails and solar arrays, are structures in which rigidity requirements are low and dimensional precision requirements can typically be satisfied without machining. Example dynamics experiments are discussed below.

6.4.1.1 SOLAR ARRAY FLIGHT EXPERIMENT (SAFE)

SAFE was launched aboard the Space Shuttle *Discovery* on August 30, 1984 as part of mission STS-41D. NASA mission archives state, "The 102-foot-tall, 13-foot-wide Office of Application and Space Technology (OAST-1) solar wing extended from the payload bay. The wing carried different types of solar cells and extended to its full height several times. It demonstrated large lightweight solar arrays for a future in building large facilities in space such as a space station [21]." The OAST-1 array was actually used for four separate experiments: SAFE, the Solar Array Flight Dynamics Experiment (SAFDE), a photogrammetric experiment, and a solar cell calibration experiment. SAFDE and the photogrammetric experiment both sought to characterize the dynamic behavior of the array when fully extended from the shuttle's cargo bay, as shown in Fig. 6.2.

Initially, OAST-1 was going to be investigated only statically. But dynamic tests were eventually added because, "In past vehicles and on-orbit structures, the structural natural frequencies were significantly higher than the control system natural frequencies. However, with large space structures, the structural natural frequencies are so low that the control natural frequency will either be very close to the first natural structural frequency or nested between a pair of the lower natural structural frequencies [22]." Twenty-three retro-reflective targets were attached to the array. The requirements for the dynamic metrology system (aside from surviving the launch) were to simultaneously track all of the targets to 19 arc seconds accuracy at up to ± 45 cm displacement. A laser diode illuminated the targets, which were recorded by the camera, both located at the base of the array. The intensity of the diode and the size of the targets were non-uniform, with greater diode intensity and larger targets as distance from the array base increased. This approach yielded images of nearly uniform target size and intensity along the length of the array. The reaction control thrusters on the shuttle provided dynamic excitation, and a time-domain algorithm was used to curve-fit the tracked target data. Four modes were identified in this manner. The identified mode shapes very closely matched the predicted shapes, but there was some discrepancy between the predicted and measured frequencies [22–26].

Fig. 6.2 NASA OAST-1 solar array extended from shuttle cargo bay [22].

Other large spacecraft structure dynamic test performed in the 1980s involved ground-based apertures. These tests were initially feasibility tests and later progressed to analyze the dynamic characteristics of the structures. One such test article was the 15 m hoop-column array, which consisted of a mesh aperture that folded up like an umbrella for storage. Several tests were performed under vacuum and ambient conditions to validate early linear static and dynamic finite element models (FEMs). It was found that effective modulus and shape information from the static tests could be used to update the dynamic models and achieve better correlation [23, 27, 28].

6.4.1.2 PHOTOGRAMMETRIC APPENDAGE STRUCTURAL DYNAMICS EXPERIMENT (PASDE)

By late 1995, NASA had abandoned the American-only space station concepts promoted during the 1980s in favor of a globally cooperative International Space Station (ISS). To gain practical experience in the long-duration missions and on-orbit docking and assembly that ISS would require, and to build international goodwill, NASA began a series of missions in which the Space Shuttle rendezvoused with the Russian Federal Space Agency Mir Space Station. The STS-74 *Atlantis* mission, launched on November 12, 1995, was the second such mission, and NASA again studied the dynamic behavior of large, lightweight, and flexible solar arrays. Instead of the OAST-1 array that was deployed directly from the shuttle cargo bay in the SAFE experiments a decade earlier, this mission would study previously installed, functioning arrays on Mir.

The STS-74/MIR PASDE was intended to demonstrate that, building on the success of SAFDE, photogrammetric measurement of articulating, non-optically targeted, flexible solar arrays and appendages was a viable, low-cost measurement option for ISS. "The Photogrammetric Appendage Structural Dynamics Experiment (PASDE) was developed to mitigate the technical risk and cost associated with on-orbit measurement of solar array and other flexible appendage structural responses for the ISS program. The experiment objectives were to demonstrate photogrammetric structural response measurement methods for solar arrays using video cameras without the use of optical targets, to provide engineering data on solar array designs similar to those expected to be used on the ISS, and to verify that routine on-orbit spacecraft operational events provide sufficient excitation for structural response testing [29]."

PASDE was part of the planning for ISS and was designed to test the effectiveness of photogrammetry as a metrology system providing the needed on-orbit data required to validate numerical models. In space, dynamic data were normally collected by strain gauges or accelerometers, signal conditioned, and transmitted to Earth for analysis using ERA or other modal parameter estimation techniques. While those same techniques were planned for use on truss elements, the expense and added mass of the wires, power, signal conditioning, electronics, and so on was considered to be too expensive and impractical on future ISS solar arrays that were 73 m long. The experimental equipment required for PASDE consisted of only three small Hitchhiker canisters installed in the shuttle cargo bay. These canisters pointed up at the Kvant-II solar array on Mir, as shown in Fig. 6.3. The canisters are denoted by red cylinders inside the shuttle bay. Each canister contained two video cameras, one focused at the root of the solar array and one focused at the tip, and each camera could record at 30 frames per second for 115 minutes.

Preparation for PASDE involved assembling and testing the Hitchhiker canisters and calibrating the cameras on the ground prior to launch. During the mission, five maneuvers were performed by *Mir* and *Atlantis* to excite in- and out-of-plane motion in the solar array. Additionally, the cameras recorded the behavior of the array during other mission operations such as docking, day-to-night terminators, and attitude maneuvers. Following the return of the shuttle, the cameras were re-calibrated.

The Russian-built array was constructed without the retro-reflective targets used in optical tracking on OAST-1, so other methods of defining points were used. Line-tracking involved tracing straight edges in the images. The intersection of two of these lines was considered a target that could be tracked through the image sequences. A cross-correlation method was also used. This method tracked the movement of a target using multiple cameras to determine its location. The photogrammetry collinarity equations were used to triangulate the three-dimensional locations of the targets in space, and these locations were tracked through the image sequences to produce time histories of all the points.

An ERA analysis was then run on the gathered temporal data to extract structural natural frequencies, mode shapes, and damping of the array. Five modes were identified between 0.1 and 0.5 Hz: two global modes of the Mir/Shuttle system and three flexible body modes of the arrays, including two out-of-plane modes and a single in-plane mode. These results illustrate the effectiveness of

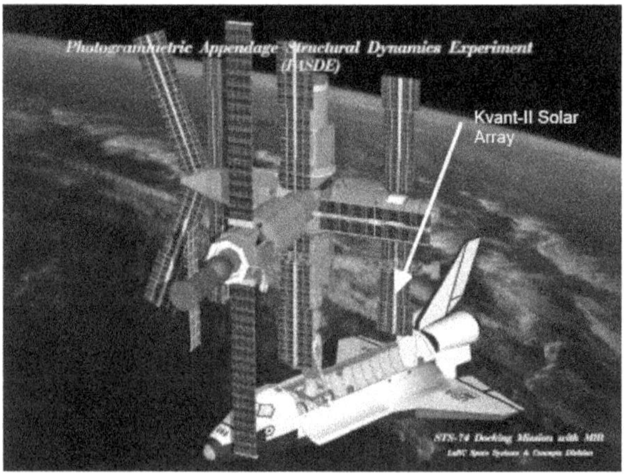

Fig. 6.3 STS-74 Shuttle/Mir configuration [29].

both photogrammetry and ERA in flight testing, as well as other time-domain techniques in the characterization of large, flexible structures. The measured array exhibited five very closely spaced modes that time-domain curve fitting techniques are particularly good at distinguishing. The photogrammetry data generated without attached targets or measurement devices from three small self-contained canisters were full field, accurate, three dimensional, and sensitive, as most of the motion captured was less than 2.54 cm peak to peak [29, 30].

6.4.2 MODERATE DIMENSIONAL PRECISION

Early testing of what would be considered modern LULS structures began in the late 1990s. These structures were comprised of ultra-thin membranes and inflatable rigidizable composite booms and struts, and dynamic data were required to validate numerical models. "Due to the uniqueness of the structures being tested, experiments conducted to date generally have required as much effort in developing the test methods as in acquiring test results. Test procedures used for traditional aerospace structures are often not applicable [31]." Therefore new dynamic testing procedures were developed in some of the following pathfinder experiments.

From approximately 1999 to 2003, the Jet Propulsion Laboratory (JPL) and NASA Langley Research Center (LaRC) conducted a series of tests to develop a database of the structural characteristics of inflatable rigidizable tubes. Several dozen cylindrical hollow tubes were tested statically and dynamically, most with length-over-diameter ratios of 10, and several much higher. Tube diameters were all between 10.16 and 15.24 cm (4 and 6 in.). Dynamic testing was conducted under fixed-free boundary conditions using impact hammers and miniature accelerometers to determine the effective elastic and shear moduli of the tubes. Like the stiff, ultra-lightweight panels tested in the 6.5.2 section, later in this chapter, the rigidized tubes were sufficiently stiff as to be amenable to this type of impact testing [31–33].

6.5 RECENT MODAL TESTING OF LULS

INTRODUCTION

Recent modal testing of LULS has included a variety of new and challenging structures which simultaneously present the possibility for substantial performance improvements over previous generations of structures and new ground and space testing challenges.

6.5.1 LEAST DIMENSIONAL PRECISION

The most common examples of low dimensional precision ultra-lightweight spacecraft structures are simple, flat, tensioned membranes. Such structures were dynamically tested in the form of scale models of the Next Generation Space Telescope (later renamed the James Webb Space Telescope) sunshield. A 1.5×1.3 m 1/10 scale model of the sunshield consisted of four layered membranes 13 microns (0.5 mil) thick. The model was suspended from the top of a vacuum chamber by a long metal rod that was excited by a shaker both inside and outside the chamber, depending on chamber size. The resulting motion of the membranes was recorded by a laser vibrometer outside of the chamber measuring through a glass porthole in the chamber wall. Deployment tests of scale models up to half the size of the actual structure were also conducted [31, 34, 35].

Many other dynamic tests of membrane-based and inflatable rigidizable LULS structures have been undertaken. References [36, 37] compared modal testing of a single quadrant of a four-quadrant square solar sail 2 m on a side in ambient and vacuum conditions. The quadrants were manufactured from 25.4-micron-thick aluminized Kapton®, and small retro-reflective targets were attached to their surfaces. In these experiments, non-contact measurement (laser vibrometer) and non-contact excitation were used. A shaker was the excitation source, but instead of being attached directly to a boom tip via a stinger, a small permanent magnet was attached to the end of the stinger, very close to, but not in contact with, the boom tip. In this manner, the boom could be excited without the parasitic mass or stiffness of the stinger and shaker. Ambient results showed several structural modes of the booms, and only one membrane mode, whereas the vacuum results showed multiple, closely spaced, low-frequency membrane modes. These low-frequency membrane modes were not seen under ambient conditions due to the apparent mass of the air surrounding the ultra-thin membranes, thus illustrating one of the key challenges of testing LULS structures on Earth as opposed to in space. And these tests were on very small, 4 m^2 test articles. The difficulty of apparent air mass only magnifies as test articles are increased toward full-scale structures hundreds or thousands of square meters in size.

Frequencies and mode shapes of partially wrinkled or creased membrane space structure ground test articles were modeled numerically and compared to experimentally determined using non-contact measurements and physically attached shakers in Refs. [38, 39].

6.5.2 MODERATE DIMENSIONAL PRECISION

To demonstrate that excitation could be actively controlled using embedded piezoactuators, several Macro Fiber Composite (MFC) actuators were embedded in composite inflatable rigidizable tubes identical to those described in the 6.5.1 section, earlier in this chapter. A laser vibrometer was used to measure the free decay vibration of the tubes from impacts. It was found that active MFC actuators increase the damping of the first bending mode from 0.4 percent to 8 percent [31, 40, 41].

The 5-m-diameter parabolic reflector in Ref. [42] was tested statically using photogrammetry of attached retro-reflective targets to determine the accuracy of the reflector shape. Development was undertaken, subsequent to the successful static characterization of the reflector shape, to perform the photogrammetry triangulation over sequences of images, generating a set of time histories of three-dimensional points. This process of applying photogrammetry to video sequences was eventually called *videogrammetry* [31, 42]. To baseline the videogrammetry measurements, dynamic data were collected at each of the retro-reflective targets using a laser vibrometer. The structure was excited by an electrodynamic shaker attached to the aluminum end caps of the composite support struts [43].

Several other membrane-based, torus-supported apertures in the form of solar thermal concentrators and antennas were dynamically tested at NASA Marshall Space Flight Center using laser vibrometry. Most of these were tested in fixed-free boundary conditions in which the metal end caps of the composite rigidizable support struts were mechanically secured to the ground. Shakers were attached to the metal end caps to provide excitation. A laser vibrometer was selected as the measurement device in these tests because of the non-contact nature of the measurements. The major drawback of the laser vibrometer is that it is single point, not full-field [31, 44–48].

The tests described in Refs. 50–57 used adaptations of various contact and non-contact modal testing methods to characterize pressurized tori. In Refs. [49, 50], a 1.8-m-diameter pressurized Kapton® torus was tested at different internal pressures using an impact hammer modified to spread the applied impulse out over a wider area by attaching a small aluminum plate to the hammer head instead of a traditional tip, and a small accelerometer to measure its response. Excitation using a tip would have caused local buckling. The first three in- and out-of-plane mode shapes were observed at pressures above 0.8 psig, but despite the modifications to the hammer, it was unable to globally distribute the excitation energy, leading to poor coherence.

A similar 1.8-m-diameter torus comprised of a 0.15-m-diameter Kapton® tube pressurized to 0.5 psi was in Refs. [51–55]. In these tests, the piezoactuator MFC patches discussed earlier were used as both the actuators and sensors. Multiple MFC patches were attached to the surface of the torus at various intervals from each other, and structural natural frequencies, out-of-plane mode shapes, and damping were obtained from the data. These tests were also performed with a membrane mirror attached to the torus to determine the boundary interactions of the two structures. Finally, in addition to excitation and measurement, the MFC actuators were used to control the structure, obtaining up to a 70 percent reduction in the amplitude of vibrations.

Similar to the other tori described earlier, the test article used in Ref. [56], shown in Fig. 6.4, has a 1.8 m ring diameter and a 0.20 m tube diameter. It is unique, however, in that it is self-supporting under ambient gravity loading, meaning that the structure is not pressurized. A regular pattern of convex hexagonal domes 8 mm side-to-side and 3.5 mm high is formed into the Kapton® membrane from which the torus is constructed. These domes give the torus its

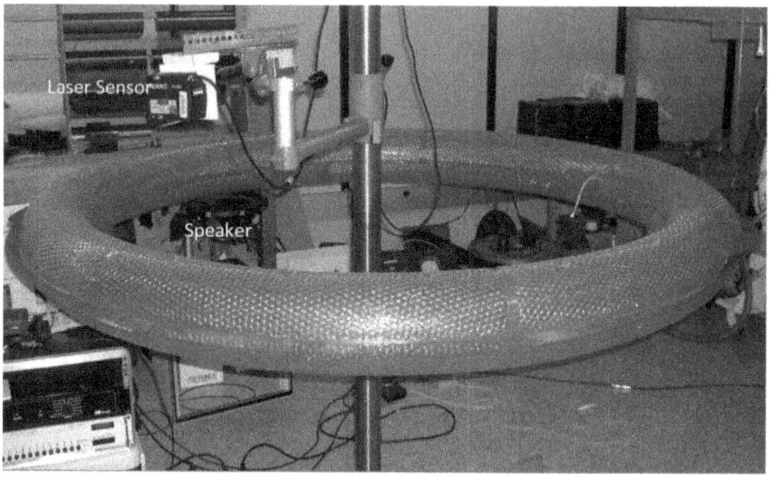

Fig. 6.4 Self-stiffening torus test article [56].

knurled appearance and provide the stiffness that enables the structure to be self-supporting.

Acoustic excitation and an impact hammer were used as the excitation methods for the different shapes of thermal-formed thin film polyimide panels tested and modeled in Ref. [57]. Given the relative stiffness of these test articles, traditional ERA analysis produced FRFs at dozens of points across the test articles, structural natural frequencies and mode shapes were extracted, and the experimental modal data were used to tune and validate FEMs. The extracted mode shapes correlated to plate and shell bending and twisting modes, in contrast to the membrane modes that will be analyzed later in this chapter. These types of thermal-formed membrane structures are therefore structurally unique in the structural dynamics testing challenges they present [57].

6.5.3 EXTREME DIMENSIONAL PRECISION

Reference [58] describes the modal testing of a flat, specular LULS membrane suspended in a hexapod comprised of rigidizable graphite-epoxy composite tubes. A *hexapod* is a circular 3.8-m-diameter torus from which the membrane is suspended, supported by six struts. The hexapod test article was suspended vertically, excited by a shaker attached to a hard plastic end cap of one of the tubes comprising the torus, and the structure and membrane were measured with a laser vibrometer. Several structural natural frequencies and mode shapes occurring between 1 and 300 Hz were identified using ERA and yielded a complete picture of the dynamic behavior of the entire structure, including membrane, torus, and struts.

Reference [59] describes videogrammetry testing to measure the in-plane dynamic behavior of a small optical-quality membrane. A 12.7-μm-thick polyester film coated with 0.2 μm of aluminum was attached with silicone adhesive to a 2-mm-thick copper ring with a 76.45 mm (3.01 in.) inner diameter and a 117.7 mm (4.64 in.) outer diameter. A Krypton Fluoride Excimer laser etched a series of 0.25-mm-diameter circular dots on the surface of the membrane by removing 5 percent of the aluminum coating. Removal of this small amount of coating did not impact the dynamic response of the membrane but was visible to the naked eye. Therefore, attaching or projecting targets onto the membrane was not required. The copper ring to which the etched membrane was adhered was attached to a shaker that provided in-plane excitation. Videogrammetry was used to track the motion of the etched targets (0.25-mm-diameter circular dots), and plotting the distance between two points on the membrane as a function of time revealed several in-plane structural natural frequencies of the membrane.

6.6 GROUND TESTING OF ATK SOLAR SAIL SYSTEM IN AMBIENT AND TVAC

The NASA In-Space Propulsion (ISP) program of the early 2000s sponsored LULS system design development and hardware demonstration activities of solar sail technology, including ambient and vacuum testing of 10- and 20-m systems, to enable or enhance a variety of road-mapped space science missions [60]. The difficulty of validating finite element analysis (FEA) modeling through testing in a fully representative environment before flight has thus far discouraged near-term mission planners from using this enabling propulsion technology. Validation of very high-fidelity models was accomplished on the 10- and 20-m system solar sail system discussed here. This section reviews descriptions of the evolution of the design, results of functional testing, and results of dynamic testing.

Large solar sail designs, one of which is shown in Fig. 6.5, call for square, four-quadrant sails 100 to 200 m on a side, 7 μm thick, with areal densities on the order of 5 g/m^2. The large size and low mass of the sail are required to generate positive acceleration from the small propulsive force provided by sunlight. For comparison, standard white paper is 100 μm thick and has an areal density of 75 g/m^2.

In the final phase of the ISP program, a larger and more complete sail system was designed, fabricated, and demonstrated, first in ambient conditions at ATK in Goleta, CA (ATK-Goleta) and later at a large vacuum-chamber facility in April and May of 2005. The 20-m solar sail hardware represents a flightworthy full (four sails, four masts) system that includes a boom for offsetting instruments well away from the plane of the sail, a dual-purposed mechanism for deployment and attitude control, tie-down and release hardware, solar panels, and launch vehicle interfaces, which were integrated in a carbon-composite central assembly that also functions as a bus chassis (shown in Fig. 6.6) [60, 62].

The 20 m ground system demonstration and validation accomplishments were planned to systematically reduce the risk of spaceflight implementation

Fig. 6.5 Four-quadrant solar sail concept [61].

through building increasingly more complex solar sail systems together with testing of the highest fidelity possible in the terrestrial environment. The progression of major hardware builds over the program is depicted in Fig. 6.7. Incremental design and demonstration were pursued to minimize cost and risk while methodically elevating the technology readiness level (TRL) of dependent systems before full-system demonstration.

After completion of the mast and sail subassembly validations shown in Fig. 6.7, the integrated 10-m quadrant was prepared for testing. To validate deployment characteristics, as well as sail shape and system dynamics, a series of tests were planned using a vacuum chamber facility at the NASA LaRC. Testing in a vacuum is necessary because the dynamics of the deploying and

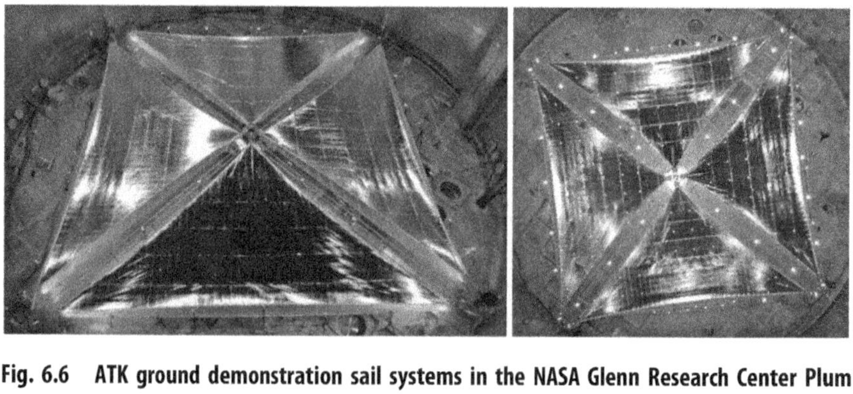

Fig. 6.6 ATK ground demonstration sail systems in the NASA Glenn Research Center Plum Brook space environment simulation chamber [63, 64].

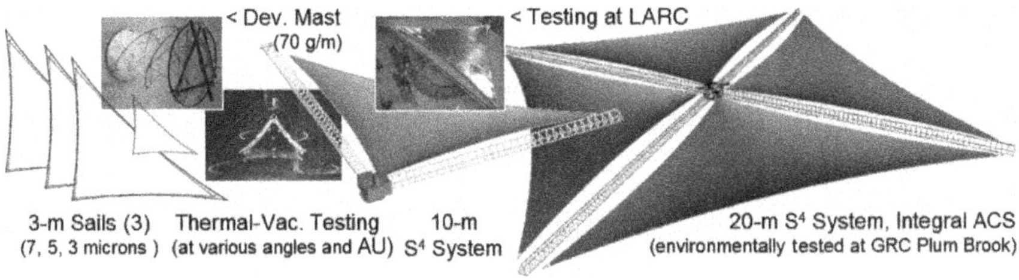

Fig. 6.7 Solar sail program milestones: progressive growth in technology maturation, size, and complexity [60].

deployed sails would otherwise be greatly affected by the surrounding air mass. The air within 2 mm of either side of the sail surface is alone equal in mass to a 3-μm-thick sail film. All sail-dynamics testing at LaRC took place under vacuum at pressures near 1 torr. The priorities for testing were, first, to validate the deployment; second, to measure the deployed shape of the (horizontal) sail billowed under gravity loading; and, third, to validate the models used to predict the system dynamics.

The test team planned an extensive series of tests to capture the data needed to support these priorities, as well as to meet goals for developing test methods applicable to 20-m-sail testing and to in-flight investigations. The testing sequence and findings are reported in detail in Refs. [62, 63]. With the system deployed in a horizontal configuration, measurement of the shape of the surface was undertaken. Sail shape is critical for two reasons. First, global shape affects torques (in-flight) that are caused by an offset between the center of pressure, *cp*, and the center of mass, *cm*. In a gravity environment, the billowed shape (1/2-m-overall gravity-induced sag) is deep enough to affect the modal response of the sails. The ability to confidently predict approximate frequencies of the first few system modes is essential for control system design needed for flight. Second, fine-scale topology affects the overall thrust performance of the solar sail. A laser radar-scanner system was employed to provide measurement of the entire sail surface with an accuracy better than 1 mm. Comparisons of the measured and predicted global billow shapes were in reasonable agreement, typically within 3 cm across the 10-m sail.

Mast dynamic testing was performed with excitation input at the sixth bay. Using laser vibrometer targets at the mast tip, the first in-plane and out-of-plane bending vibration modes and the first torsional vibration mode were captured. Greater mass distributed along the mast (harnessing and vibrometry targets) and at the tip, in combination with the central structure and ground support equipment (GSE) flexibility, resulted in lower mast frequencies (4.4 vs 5.7 Hz) than were recorded during earlier component-level testing at ABLE. In addition, the frequencies were slightly lower than predicted (4.7 Hz) for the integrated masts. The first four modes of the system were captured in the test, and overall the agreement between the predicted and measured shapes was reasonable. The first-mode frequency prediction by FEA was lower than that found in the test. This discrepancy could be expected given that the FEA overpredicted the billow depth (likely due to halyard tension uncertainty and load angle variation), which can account for an underprediction of the first-mode frequency. The higher-frequency mode shapes measured were as predicted. The frequency predictions were generally higher because they are more dependent on mesh density, the effective extensional modulus, and the stress distribution, which are interrelated in the prediction of local shape and stiffness. The soundness of the fundamental elements of a deployable integrated sail and structure were validated with the tests performed on the 10-m quadrant.

The evolved 20-m solar sail system integrates the gossamer coilable mast and sail membrane technology, solar arrays, launch tie-down and release mechanisms, and attitude-control actuators. Each of these items is efficiently packaged within structure shared by other bus components and mission payloads. These subsystems together form a generic scalable solar sail possessing reliable deployment, structural robustness, and determinate sail shape with, most critically, minimized overall mass and volume. The mast structures used on the 20-m system were "truncated" in length, meaning that the diameter (40 cm) was sized for the needs of an 80-m spaceflight system. Because the system "size" is 20 m from mast tip to mast tip, approximately 14-m-long masts were required. The central assembly, approximately 1 m across, was sized to hold enough mast structure for a 40-m sail system. The sizing choices for the central assembly and masts were made to allow the maximum benefit from design, fabrication, and validation activities in the ground demonstration toward a flight validation of a 40-m- class solar sail.

A ground-controllable pan-tilt camera was ideal for tracking particular areas of interest such as boom uncoiling, lanyard advancement, tip kick-over, ballast movement, spreader-bar motion, halyard mechanization, sail synchronization, and inspection of deployed-sail surface morphology. Documentation of the deployment process was of primary interest, followed by identification of the sail shape and the fundamental system modes. Because the sail was nominally flat and the targets were applied at known locations, the verification of a nominal global, billowed shape was accomplished with photogrammetric analysis of images from a single camera location. The fundamental system mode shapes could have been captured with videogrammetry but would have required high camera resolution and large targets, altering the local stiffness and mass of the membrane. Because of the small displacements involved, a more practical instrumentation option was accelerometers. The 20-m system demonstrated the inclusion and operation of a suite of accelerometers, a host of other embedded systems, and a camera platform, for full characterization and in-flight behavioral identification.

The 20-m solar sail system is shown in Fig. 6.6, as oriented inside the Plum Brook vacuum chamber in the deployed test configuration. Retro-reflective targets on the sails, mast tips, spreader bars, and various scale bars (used for photogrammetry) are also visible. Baseline sail dynamics testing at Plum Brook took place at pressures of 1 – 2 torr. Measurements of dynamic response were made primarily at the 44 retro-reflective targets on the sails, using a laser vibrometer. Baseline testing used dynamic excitation by magnetic forces (non-contact) at the quadrant corners. Other excitation methods were evaluated for their suitability to on-orbit modal investigations, such as piezoelectric stack actuation from the mast roots, spreader-bar rotation (at the mast tips), and bimorph piezoelectric patch actuation on the sail cords. The masts were tested separately, using piezostack actuators (in ambient atmospheric conditions) to provide a baseline for the response of these primary structural members. This approach could be followed in-flight as well, if sufficient time is arranged before the deployment of the sails.

The first two modes were found to agree within 1 percent on all four masts. The frequency responses for the Z-axis mast are shown in Fig. 6.8a. The first mode, at 0.81 Hz, represents cantilever bending in which the tip moves in-plane, with slight pendulum effect stiffening from the 1.8-m overhead support cable. The second mode is driven by mast torsional compliance and the tip-mass rotational inertia. The third peak in the FRF is attributed to the tip mass bouncing vertically on the support cable. The fourth peak corresponds to the second in-plane bending mode. The response of all four masts during system testing is shown in Fig. 6.8b.

The response of all four masts during system testing was also measured. The modes are lower because the masts are supporting the weight of the sails. For system testing, the changes in tip offloading consist of the addition of a negator spring in line with the support cable (at the top) and a 2.3-kg counterweight mass added to the tip plate. The negator spring system provided a low-frequency suspension to free up system modes that are out-of-plane. The counterweight mass was cantilevered one-half meter above the tip plate, to counteract the torsion imparted by the sail halyards when the spreader bar is rotated away from the 0-deg (balanced) position. Increases to the effective rotational inertia at the tip cause the first twist mode to drop to 0.50 Hz. The second FRF peak was a combined twisting and bending response at 1.06 Hz. The in-plane mode was not captured because the reflector aligned to register mast lateral motion was not captured in system testing. The observed modes agree to within 2 percent relative to the pretest predictions. Excitation responses of the masts with the piezostack actuators produced clean response peaks (note that the data in Fig. 6.8 are plotted in a semilog format). This mode of excitation was originally developed for the SAILMAST flight experiment [60].

Measurements of system dynamics were made for various magnet-excitation combinations and input parameters, at both the quadrant level and system level. Bending and torsion of the masts dominate the lowest modes of interest on the 20-m system. This response is due mostly to the presence of the eccentric counterweights mounted to the mast-tip plates. The counterweight reacts to the torque induced on a mast that develops from the tensions of the opposing sail halyards as a spreader bar is tilted. This mass had to be used when the spreader bar is at 0 deg as well, because the offloading negator provides tip support to only the

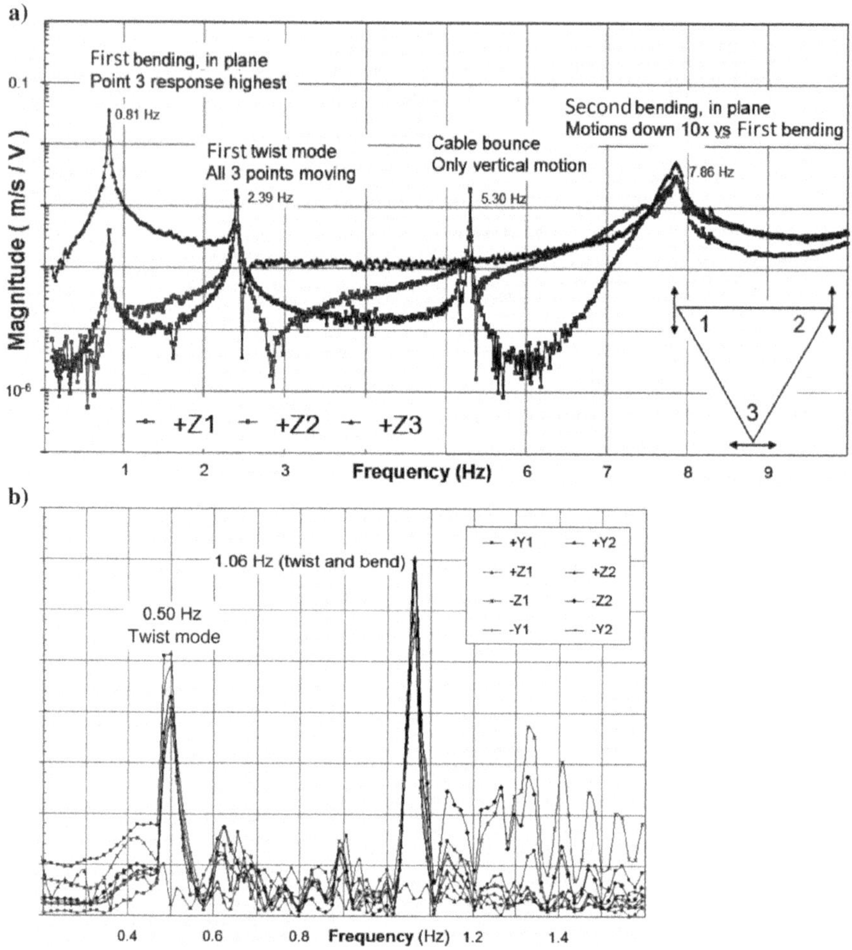

Fig. 6.8 Mast tip responses a) in ambient conditions and b) with sails and counterweights in vacuum [60].

nominal weight. For the baseline test configuration, the sail spreader bars were horizontal and then they were rotated 22.5 deg for another sequence of tests. Similar modal responses were seen in the four sails at each position. The range of expected modes and test findings for the baseline sail at the nominal spreader-bar orientation (0 deg) are shown in Fig. 6.9. Listed in the gray bar regions aligned along the graph of target responses are the dominant expected mode frequency ranges. The responses that fell below the response of the target beside the excitation input (bold line) were trimmed out for clarity. The first grouping of peaks indicates fundamental modes in the range of 0.50 and 0.62 Hz. Each of the sails has a slightly different mass, and so this creates sets of four spread low-frequency responses, one for each sail's fundamental mode. As shown in the inset to the graph in Fig. 6.9, the lowest mode responses for magnet 2 (M2) excitation were similar to those of the magnet 1 (M1) excitation results.

The shape predicted by FEA corresponding to a fundamental mode of Q4 at a frequency of 0.483 Hz is shown to the right of the graph in Fig. 6.9, above the ODS found in the testing. The same response at 0.50 Hz was found for various inputs (i.e., at M1 and at M2) and for out-of-phase excitation of M1 and M2. The second mode, shown to the right in the figure, corresponds to mast bending. Again, there were four related modes predicted from 0.62 Hz to 0.64 Hz for various combinations of mast motion phasing. The modal responses found in the test peaked at 0.625 Hz for the three input conditions, but the ODS recorded for each varied slightly. As can be seen in the three quadrant details in Fig. 6.9, the nodal line shifted toward the input, as may be expected in a very flexible nonlinear system.

The best location for modal excitation is, ideally, at the antinodes of the system response. Magnets were positioned at the three corners of the sail and at the middle of the hypotenuse cord. The latter was unusable due to variations in the sail position after pump-down. Condensation formed on the membrane as the chamber was initially evacuated, increasing the effective mass of the sail, and the negator-spring-mounted halyards were extended. When the chamber reached a substantial vacuum, the water evaporated, returning the supported load to a nominal value. A small amount of positional variation (due to hysteresis) in the negator mechanism was amplified to result in a positional variation of 5 to 15 cm at the midpoint of the sail hypotenuse. This variation was greater than the stroke of the magnet-position adjustment actuators, and the testing schedule did not allow time for diagnosis, mitigation testing, and added pump-down days.

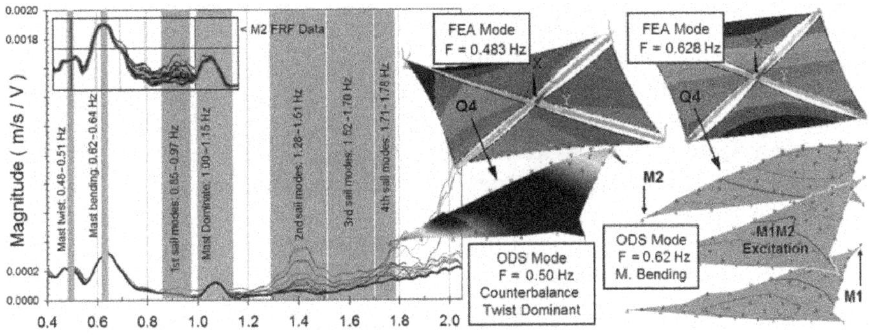

Fig. 6.9 System dynamic responses to M1 input, lowest modes due to mast-tip counterweight [60].

The ground demonstrations and validations of the ATK solar sail system technology elevated the Technology Readiness Level (TRL) of solar sailing in all areas critical to support a flight validation mission in earth orbit. NASA's In-Space Propulsion Technology Program systematically reduced the risk of flight implementation with substantial technology development through a series of demonstrations of increasingly more complex solar sail systems and by testing at the highest fidelity possible in the terrestrial environment. The progressive demonstrations at the component, 10-m, and 20-m levels elevated the TRL of solar sailing technology in all critical areas. The soundness of the design for deployment and control of a fully integrated sail and structure system has been validated through an extensive series of tests. Analytical correlation activities showed sound agreement with test results. The continuum of system design space and analytical modeling capabilities together ensure a functionally scalable design and mathematically predictable result for larger systems such as a flight demonstration system at 40-m scale and mission systems of up to 200 m on a side. References [60, 62–65] have full details.

6.7 MEGAFLEX AND ULTRAFLEX

With the launch of Deep Space 1 (DS1) in 1998, NASA began a new era of possibilities for exploring the solar system. For 16,000 hours, a revolutionary ion engine powered by a concentrator solar array built by Orbital ATK—Goleta (OAG) ejected xenon ions at 25 miles per second, slowly adding acceleration with a mass efficiency 10 times that of conventional rocket engines [66]. DS1, with solar arrays that spread 48 feet tip-to-tip to produce 2500 watts [67], traveled to 1.6 astronomical units AU and caught close-up views of a comet. NASA took a second leap forward in solar electric propulsion in 2007, with the launch of the Dawn spacecraft built by Orbital ATK. The three ion engines on Dawn, together with a 10-kW solar array, have propelled the spacecraft to targets in the asteroid belt (as far at 3 AU from the sun) [68], with a net velocity change of over 10 km/s—far more than any other spacecraft has been able to using conventional chemical engines.

Larger ion engines—and larger solar array systems to power them—are key elements for America's future in space. NASA's strategic roadmaps for exploration, science, and advanced technology all consider solar electric propulsion (SEP) an essential capability to develop further, enabling a host of advanced deep space missions.

Many NASA missions road-mapped by the Science Mission Directorate (SMD) and documented in the most recent decadal survey [69] plan to employ SEP. UltraFlex wings are baselined as the power-generation system for the majority of these missions due to the high performance and flight heritage of the technology. The Human Exploration and Operations Mission Directorate (HEO) plans for much larger SEP architectures—with massive solar array systems—that will one day help support a sustainable, affordable human presence in space [70].

The NASA Game Changing Development (GCD) program targeted development of new solar array systems capable of supporting SEP applications at power levels from 25 kW to 250 kW. OAG recognized the scaling potential of UltraFlex [71] and sought to build upon and extend the heritage UltraFlex technology to meet the challenges put forth by the GCD program. The Phase I effort succeeded in maturing the MegaFlex solar array system concept to TRL 5+ in support of future 30- to 50-kW SEP applications while being challenged to develop analytical methods to scale the concept to a total system power level of 250 kW or greater. The technology development effort successfully produced an integrated design for all electrical and mechanical systems, demonstrated with a flight-like 10-m-diameter wing.

MegaFlex is simply UltraFlex technology enhanced to enable very high-power applications through the introduction of two new features: folding spar joints and panel extension hinges. These features allow stowage in an even more compact volume with the longest-length dimension significantly reduced. For example, with the spars folded at one-third out from the center, a 30-m-diameter wing will stow for launch with a major dimension of 10 m, which is compatible with the spacecraft anticipated for such a mission. Two wings of this size can provide 450 kW, and they will package in just a small fraction of the fairing volume of launch vehicles under development. The potential for power will rise as cell efficiency continues to improve in the coming years. But already, a two-wing MegaFlex array can provide well beyond the 250-kW power-level challenge levied by NASA and be easily stowed in a 4-m launch fairing . An overview of the structure deployment is shown in Fig. 6.10, and full details can be found in Ref. [70].

The wing hardware validated the parametric MegaFlex mass model by correlating very well to the predicted 10-m wing mass. The MegaFlex Demonstration Unit (MDU) demonstrated compliance to the program mass requirement, with performance that exceeded 100 W/kg at the end of the (modeled) mission life, when the cells have accumulated a very high radiation dose, while simultaneously meeting the challenging strength and stiffness requirements.

High-fidelity analyses were performed to evaluate the stowed and deployed mechanical and thermal behaviors of the wing. The stowed system behavior and the structural capability with respect to the defined launch requirements were investigated with models that were evolved and refined over the course of the program as the design developed and as test data on components and subassemblies became available. The modeling was employed to support and guide the design and to predict behaviors such as stiffness, mode shapes, frequency, and stress for the acoustic and random vibration environments of launch.

Verification of the modeling and the robustness of the hardware were demonstrated by testing to acoustic levels enveloping Atlas and Delta environments. To validate survival to the random environment, the acoustic input was increased until accelerometers at key points responded with equal or higher root mean

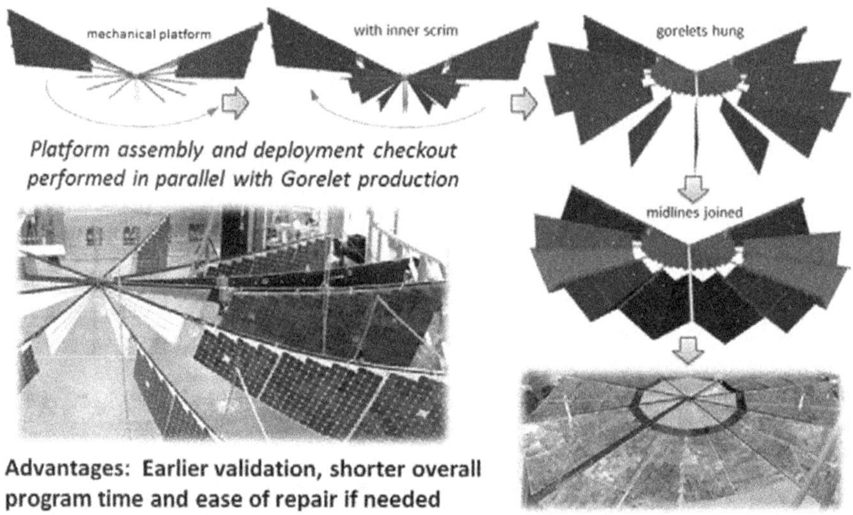

Fig. 6.10 Parallel flow in MegaFlex platform and power assembly [70].

squared (RMS) deflection than forecast by the FEM. The lowest mode predicted for the stowed wing, with refinements to reflect the flight design conditions, was 28 Hz, meeting the requirement of a minimum of 25 Hz.

The deployed behavior and capability to perform against the defined structural requirements for both in-flight (0.1 *gg* max) and 1-*g* environments were considered, supported where needed by detailed component modeling using FEA, Mathcad, and Excel computations. The deployed FEA model was studied using nonlinear large deflection analyses to arrive at accurate gore tension and catenary deflections. The deflections and stiffness matrices from the nonlinear analysis were then used in modal analyses that provided frequency predictions, which ended up being within 3.5 percent, on average, of the first three frequencies as measured in vacuum at Plum Brook. Ground support equipment present in the offloaded (1*g*) model, but not in the 0 *g* model, included the springs and cabling associated for supporting the spars. The stiffness of these elements was individually measured and modeled. After extrapolation of the validated 1 *g* model to 0 *g* conditions, the lowest mode in flight was predicted at 0.27 Hz, which was well above the requirement of 0.1 Hz.

Typically, high-power solar array wings for geosynchronous earth orbit (GEO) missions are designed to obtain a first mode of 0.05 Hz. The MegaFlex wing was demonstrated (in vacuum) to have a deployed natural frequency greater than the required 0.1 Hz, which requires four times more stiffness. Stiffness tests were performed at the subsystem level on the extension hinge elements. Results were within 10 percent of the FEM predictions. This requirement was verified by both analysis and testing of the MDU. The measured first mode, extrapolated to 0 *g*, was 0.26 Hz, which provided a significant stiffness margin above the requirement. The strength requirement drove the structural sizing (e.g., spar cross-section), and the circular configuration efficiently utilizes the material to provide a very high deployed stiffness.

The static acceleration requirement for the stowed wing was 20 g in each axis. This requirement was confirmed by analysis that demonstrated positive margins of safety for the MDU. In addition, the wing was subjected to the dynamics of the launch environment, to levels which enveloped the maximum predicted acoustic qualification environment of 145 dB overall sound pressure level (OASPL). Following this initial acoustic test, the levels were increased until accelerometer responses matched analytical predictions for deflections in response to the random vibration qualification requirements, resulting in an OASPL of 153 dB.

The stowed frequency of the MegaFlex design was predicted as 28.4 Hz, which exceeded the required 25 Hz. The response of the accelerometers in acoustic testing could not be used to precisely identify the lowest modes.

The ground demonstration and validation of MegaFlex technology has elevated the TRL in all critical areas necessary to support a low-risk flight validation mission. Two UltraFlex arrays are deployed on Mars (NASA Phoenix lander mission, 2008), and two more will be landing there in 2019 (NASA InSight lander mission). OAG completed a 10-wing build program of 3.7-m wings that currently fly on resupply missions to the ISS.

6.8 INTEGRATED SOLAR SAIL DIAGNOSTICS SYSTEM

In support of NASA ISP, an integrated diagnostics instrumentation package capable of accurate monitoring of solar sail structures during ground and near-term flight experiments was developed. The approach taken focuses on lightweight optical sensing techniques based on photogrammetry to measure solar sail shape, and dynamics and thermography to measure its temperature. Some non-optical sensors, including accelerometers, load cells, and thermocouples, were also used. The diagnostics package must be low mass (<10 kg for scale-model sails, evolvable to 2 kg for a 5000 m^2 mission sail) and must measure/monitor key sail structural parameters such as sail support tension, boom and sail deflection, boom and sail natural frequencies, sail temperature, and sail integrity [72].

Vibration measurements on solar sails are particularly challenging due to the very low frequencies (<1 Hz) of the first several modes, extremely large sizes, large stiffness variation between the booms and sails, and modal coupling between boom and sail dynamics, among other problems. Moreover, theoretical questions remain, including, How do effects like wrinkling, thermal loading, and manufacturing variability affect the vibrational response of membrane/inflatable space structures in general, and solar sails in particular? Can adaptive control methods use vibration information for local and global state estimates in such structures? [73, 74]

Table 6.2 lists the types of measurements needed for validating structural analytical models and modeling techniques of solar sails. Also listed are estimated ranges of the measurement parameters and an accuracy target for each parameter assuming a sail size of 70 m. Accuracy targets for other sail sizes vary proportionately. This information is based in part on requirements listed for a proposed NASA New Millennium Program ST7 solar sail flight experiment [75]. Note that the desired measurement parameters contain two stress quantities. There are no known methods to directly measure stress in structures. Stress is always derived indirectly from directly measured quantities, typically kinematic quantities such as static displacement or strain, or dynamic frequency, damping, and phase. These kinematic quantities are then coupled with a model, such as material constitutive relations, to derive the associated stress. For an in-space measurement system, an attractive non-contacting method for measuring the desired kinematic quantities of sail shape (displacement) and sail dynamics (frequency, damping, and mode shapes) is photogrammetry.

In practice, the measurement accuracy achieved with photogrammetry is highly dependent on several mutually dependent factors, including

- Size and geometry of the test structure
- Number of cameras and their image resolution
- Camera synchronization (for dynamics measurements)
- Locations and pointing directions of cameras
- Illumination conditions
- Clarity and contrast of surface features (targets)
- Camera and lens characteristics
- Image compression procedures
- Camera stability
- Calibration and data analysis procedures

Numerical simulations were conducted to calculate the expected photogrammetric measurement precision of various assumed camera geometries for a square solar sail structure. Figure 6.11 shows the selected configurations. Each symbol in Fig. 6.11 indicates the location of one or more cameras, as noted. Each camera used an imaging sensor with 1024×1024 pixels, and the assumed target marking (centroiding) precision was 0.5 pixels. Figure 6.12 shows the corresponding results for in-plane measurement precision, out-of-plane measurement precision, and total mean measurement precision. Each bar in Fig. 6.12 is scaled relative to the maximum value obtained, which is the in-plane result for camera configuration 2A. Note that more-precise (better) measurements correspond to smaller bars in these plots. As expected, camera configuration 3 has the best measurement precisions, but, of course, of those studied, this design is

TABLE 6.2 MEASUREMENT GOALS FOR A 70 M SOLAR SAIL FLIGHT EXPERIMENT

Parameter	Expected Range of Values	Accuracy Target	Notes
Deployment Dynamics	Qualitative: Video coverage of the entire deploying sail Quantitative: Measure trajectories of the deploying booms and membranes	Qualitative: High-quality video at 10+ frames per second Quantitative: 40 mm	These data are measured only once, so cameras for this purpose can be located off the solar sail (e.g., on a carrier platform) or ejected afterwards to minimize mass.
Deployed Shape	0 to 2 m (est.) from planarity	20 mm	20 mm is the mean accuracy for a uniformly distributed grid of targets.
Deployed Vibration Characteristics	Natural Frequencies: First 10 system modes <1 Hz Damping: $<2\%$ Amplitudes: <1 m	Natural Frequencies: Within 1% Damping: Within 10% Amplitudes: 20 mm	Image frame rate of cameras is at least two times the highest vibration frequency of interest.
Sail Support Tension	0 to 50 N	0.1 N	To investigate *in-situ* measurement possibilities with suppliers.
Boom Stress	Deployment: $0 - 15 \times 10^6$ Pa Operational: $<1 \times 10^6$ Pa	Deployment: 10^5 Pa Operational: 10^4 Pa	The highest stress levels occur during deployment, but it is difficult to measure then. Operational stresses are much lower than deployment and are also difficult to measure directly.

(*Continued*)

TABLE 6.2 MEASUREMENT GOALS FOR A 70 M SOLAR SAIL FLIGHT EXPERIMENT (*CONTINUED*)

Parameter	Expected Range of Values	Accuracy Target	Notes
Sail Film Stress	$0-10^6$ Pa (mostly under 10^5 Pa)	10^3 Pa	Will experience very low operational stresses. Can be estimated using measured wrinkle patterns, sail shape, modes, and predictive structural model.
Sail Temperature	Full-field measurement with IR camera (thermography)	10 deg C	Is important for some solar sail missions, particularly in the inner solar system.
Sail Integrity	Identified by data trends versus time and/or direct optical observation	High-resolution imagery, perhaps on one sail quadrant only	If not mass-prohibitive, a high-resolution camera will include pan/tilt/zoom.

the most difficult to implement since it requires a camera mast at the tip of each sail boom, which may be unacceptable for a flight vehicle. The second-best measurement configuration of those studied is configuration 1, which uses a stereo pair of camera clusters on an offset, horizontal connecting bar. Configuration 4 provides metrology on both sides of the membranes.

Cables that collect loads from the sail membrane quadrants and carry them to the booms complicate the sail's structural configuration. The cables and membranes carry tensile loads, while the booms carry compressive loads and some bending. Boom loads are likely to be small (perhaps only a few newtons), as are boom deflections. Sail loads are also likely to be small, but deflections may be considerably larger than those of the booms due to the relative differences in stiffness. This wide range of stiffness, coupled with potentially large sail displacements and nonlinear material behavior, provides for challenges in predicting sail response. Boundary conditions are always difficult to know with certainty in any complicated structure. The dynamic response of membranes, due to their no-compression/wrinkling character, is extremely challenging to accurately predict. This implies that, even under the best situations, any sail diagnostics measurement package must be designed to work with considerable uncertainty with respect to structural model predictions.

An important factor in the design of a sun-side photogrammetric measurement system viewing the entire sail is reflected glare and glints from the sail. In most cases, at least one imager will see an intense specular reflection of the sun. Over much of the rest of the sail, wrinkles and creases can cause local specular reflections to the imagers. The glinting width of these features will generally be $\ll 1$ pixel, so the brightness seen by the imager will be far less than that of the light source. Full glare can be about 60,000 times brighter than a diffuse-white photogrammetry target, while the diffuse reflection from a non-glinting sail is only about 2 percent as bright as a target. Therefore, even "micro-glint" can impede image analysis. The brightness of the specular reflection of the sun will be about 85 percent of full sunlight, due to the imperfect specular reflectivity of the membrane. However, the sail will generally have some local curvature. If the membrane is concave in the region of the specular reflection, the "sunspot" focused onto the imager may be larger and may involve several times more energy than in a direct image of the sun through the same lens. The camera systems must be designed accordingly to handle the worst-case brightness condition.

"Glint" is here used to refer to specular reflection from local creases, wrinkles, or other regions smaller than a pixel (on the camera imager) in at least one direction. This glint can result in pixel brightness anywhere in about the 3,000,000:1 brightness range between the full intensity of a specular reflection and the diffuse brightness of the sail. Glint should not be a significant problem if membranes are under enough biaxial tension to remove stowage creases and to prevent wrinkles (which occur under uniaxial tension). In "suitably tensioned" sails, only seam and edge imperfections are likely to cause significant glint, and keeping photogrammetry targets far enough away from such features can

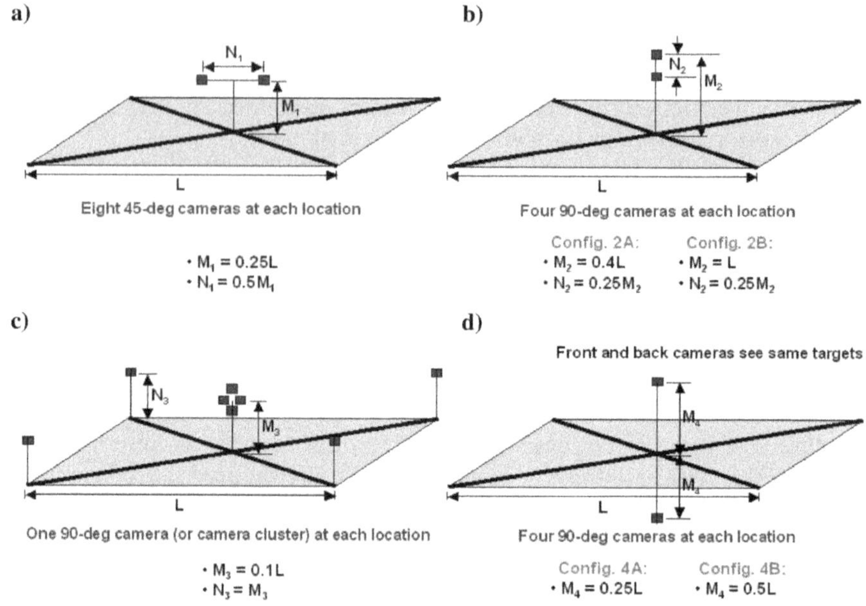

Fig. 6.11 Candidate camera configurations for photogrammetry. Symbols show camera locations [76]. a) Camera configuration 1. b) Camera configuration 2. c) Camera configuration 3. d) Camera configuration 4.

prevent image-analysis problems. However, if the sail membrane tension is kept low during and after deployment, creases from stowage may never fully straighten out. Creases may cause the glint level to be quite high, and variable across the sail. It seems unlikely that ultra-thin membranes will form creases only at the intended

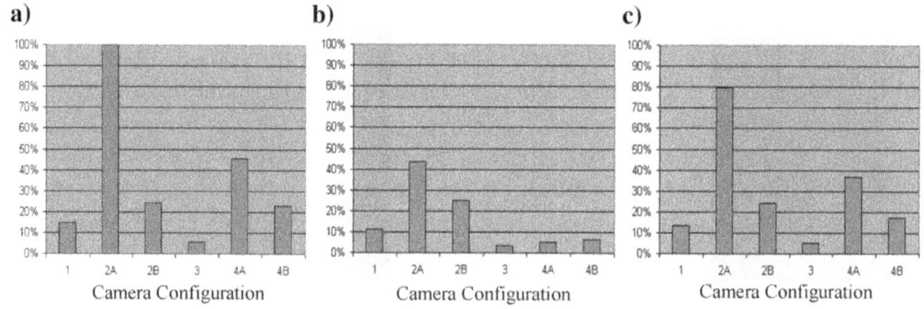

Fig. 6.12 Relative photogrammetric measurement precisions for candidate camera configurations. (Smaller is better.) [76]. a) In-plane precision. b) Out-of-plane precision. c) Total mean precision.

folds when flight-size quadrants of several thousands of square meters are folded. Pleats will probably form in random locations, especially when the once-folded sail is folded again, at right angles to the initial folds, for stowage in the launch container.

In summary, glare and glint are important design considerations for a solar sail photogrammetric measurement system for several reasons, including

- Potential permanent sensor damage from focused sunlight
- Partial or total loss of useful imaging when glare is too strong
- Difficulty finding and centroiding targets and other features when glint is highly variable
- Difficulty in evaluating long-term changes in diffuse reflectivity if glint is significant

There is one positive aspect of glint: Glint variations may themselves have diagnostic utility.

Full details of this system are available in Ref. [76].

6.9 SUMMARY

Assuming one is familiar with ground-based structural dynamics testing of traditional stiff structures [9], Table 6.1 provides a good foundation for approaching the testing of LULS structures. Key considerations include test goal, excitation method, measurement method, and structural class/dimensional precision. Important practical steps for each are as follows:

- Test goal (frequency, mode shape, damping)
  - Structural natural frequencies are dominated by the tension carried by the structure or membrane in structures that are not self-supporting in 1 g.
  - Mode shapes are very difficult to accurately measure on the ground due to the added mass and viscosity of air; deflection shapes can be substituted. Testing under vacuum is time consuming and challenging, and, given that the structures under test typically are designed to carry loads in space orders of magnitude less than 1-g, gravity loading complicates mode shape extraction even under vacuum.
  - Structural damping is almost impossible to separate from air and gravity on the ground, and it is therefore infrequently reported.
- Excitation method (contact, non-contact)
  - Almost all traditional structural dynamics excitation methods involve physically attaching to or contacting the structure under test. For LULS structures, physically attaching excitation sources alters the local mass and stiffness of the structure under test and, due to the lack of structural stiffness, tends to be insufficient for simultaneously exciting the entire

structure. Consider non-contact excitation methods, including ambient noise.

- Measurement method (contact, non-contact)
  - As is the case with the excitation method, almost all traditional structural dynamics measurement methods involve physically attaching to or contacting the structure under test. For LULS structures, attaching measurement devices, including retro-reflective targets, to the structure alters the local mass and stiffness of the structure under test. Consider non-contact measurement methods, including videogrammetry and laser vibrometry.
- Structural Class/Dimensional Precision (least, moderate, extreme)
  - The structural class affects choice of measurement type and test setup. The simplest test setups are usually the best, assuming all the data requirements can be met. Non-contact measurement techniques are capable of collecting at a variety of scales but will typically need to be recalibrated to capture the range from low-frequency, high-amplitude testing to high-frequency, low-amplitude testing.

REFERENCES

[1] Thompson, W. T., and Dahleh, M. D., *Theory of Vibration with Application*, 5th ed., Prentice Hall, Upper Saddle River, NJ, ISBN-0-13-651068-X, 1993.

[2] Ewins, D. J., *Modal Testing: Theory and Practice*, Research Studies Press Ltd., Somerset, England, ISBN 0863800173, also Wiley Inc., ISBN 0471904724, 1984.

[3] McConnell, K. G., *Vibration Testing, Theory and Practice*, John Wiley & Sons, Inc., New York, NY, ISBN 0471304352, 1995.

[4] Inman, D. J., *Vibration, With Control Measurement, and Stability*, Prentice-Hall, Inc., Englewood Cliffs, NJ, ISBN 0139427988, 1989.

[5] Allemang, R. J., *Vibrations: Experimental Modal Analysis*, University of Cincinnati Structural Dynamics Research Laboratory, March 1999, UC-SDRL-CN-20-263-663/664.

[6] Allemang, R. J., *Vibrations: Analytical and Experimental Modal Analysis*, University of Cincinnati Structural Dynamics Research Laboratory, Feb. 1999, UC-SDRL-CN-20-263-662.

[7] Allemang, R. J., *Experimental Modal Analysis, Lecture Notes, Parts I and II*, University of Cincinnati Structural Dynamics Research Laboratory, June 2005.

[8] Hardin, J. C., *Introduction to Time Series Analysis*, NASA Reference Publication 1145, Nov. 1990.

[9] "The Fundamentals of Modal Testing," The Modal Shop, Inc., Agilent Technologies, http://www.modalshop.com/techlibrary/Fundamentals of Modal Testing.pdf, 2016.

[10] Juang, J. N., and Pappa, R. S., "An Eigensystem Realization Algorithm for Modal Parameter Identification and Model Reduction," *J. Guidance, Control, and Dynamics*, Vol. 8, No. 5, Sept.–Oct. 1985, pp. 620–627.

[11] Longman, R. W., and Juang, J. N., "Recursive Form of the Eigensystem Realization Algorithm for System Identification," *J. Guidance, Control, and Dynamics*, Vol. 12, No. 5, Sept.-Oct. 1989, pp. 647-652.

[12] Juang, J., and Pappa, R. S., "A Comparative Overview of Modal Testing and System Identification for Control of Structures," *SEM Spring Conference on Experimental Mechanics*, Houston, Texas, June 1987.

[13] Black, J., "Photogrammetry and Videogrammetry Methods Development for Solar Sail Structures," NASA Contractor Report, NASA/CR-2003-212678, Sept. 2003.

[14] Pappa, R. S., Black, J. T., Blandino, J. R., Jones, T. W., Danehy, P. M., and Dorrington, A. A., "Dot Projection Photogrammetry and Videogrammetry of Gossamer Space Structures," *J. Spacecraft and Rockets*, Vol. 40, No. 6, Nov.-Dec. 2003, pp. 858-867.

[15] Black, J. T., and Pappa, R. S., "Photogrammetry and Videogrammetry Methods for Solar Sails and Other Gossamer Space Structures," 45th AIAA/ASME/ASCE/AHS/ASC Conference on Structures, Structural Dynamics and Materials, Palm Springs, CA, AIAA Paper 2004-1662, Apr. 2004.

[16] Black, J. T., and Pappa, R. S., "Videogrammetry Using Projected Circular Targets: Proof-of-Concept Test," NASA Technical Memorandum, NASA/TM-2003-212148, Feb. 2003.

[17] Jenkins, C., Gough, A., Pappa, R., Carroll, J., Blandino, J., and Miles, J., "Design Considerations for an Integrated Solar Sail Diagnostics System," 45th AIAA/ASME/ASCE/AHS/ASC Structures, Structural Dynamics and Materials Conference, Palm Springs, CA, AIAA Paper 2004-1510, April 2004.

[18] Blandino, J. R., Pappa, R. S., and Black, J. T., "Modal Identification of Membrane Structures with Videogrammetry and Laser Vibrometry," 44th AIAA/ASME/ASCE/AHS/ASC Structures, Structural Dynamics and Materials Conference, Norfolk, VA, AIAA Paper 2003-1745, April 2003.

[19] Hedgepeth, J. M., "Critical Requirements for the Design of Large Space Structures," NASA Contractor Report, NASA-CR-3484, 1981.

[20] Jenkins, C .H., and Korde, U. A., "Experimental Membrane Vibrations: A Review and New Results," *J. Sound Vibration*, Vol. 295, No. 3-5, 602-613, 2006, DOI: 10.1016/j.jsv.2006.01.036.

[21] NASA Mission Archives, STS-41D, http://www.nasa.gov/mission_pages/shuttle/shuttlemissions/archives/sts-41D.html, Oct. 2006.

[22] Schock, S. W., "Solar Array Flight Dynamic Experiment," NASA Technical Paper 2598, Document ID 19870003148, Accession Number 87N12581, May 1986.

[23] Pinson, L. D., "Recent Advances in Structural Dynamics of Large Space Structures," *NASA Technical Memorandum*, NASA-TM-100513, Oct. 1987.

[24] Cáceres, M., "Space Shuttle Era Nears Its End," *Aerospace America*, Sept. 2006, pp. 12-14.

[25] Brumfield, L., Pappa, R., Miller, J., and Adams, R., "Orbital Dynamics of the OAST-1 Solar Array Using Video Measurements," AIAA Paper 85-O758-CPy, 1985.

[26] Brumfield, L., Pappa, R., Miller, J., and Adams, R., "Orbital Dynamics of the OAST-1 Solar Array Using Video Measurements," IAF Paper 85-213, 1985.

[27] Belvin, W. K., and Edighoffer, H. H., "The 15 Meter Hoop-Column Antenna Dynamics: Test and Analysis," NASA Center for AeroSpace Information (CASI), Structures Interaction Technology, 1986, pp. 167-185.

[28] Neiswander, R. S., "Inflight Optical Measurement of Antenna Surfaces; Triangulation to Measure Deformations of Large Space Structures," *NASA, Langley Research Center Large Space Systems Technologies*, Vol. 1, Jan. 1978, pp. 457–489.

[29] Gilbert, M. G., Welch, S. S., Pappa, R. S., and Demeo, M. E., "STS-74/MIR Photogrammetric Appendage Structural Dynamics Experiment Preliminary Data Analysis," 38th AIAA/ASME/ASCE/AHS/ASC Structures, Structural Dynamics, and Materials Conference, and AIAA/ASME/AHS Adaptive Structures Forum, Kissimmee, FL, AIAA Paper 97-1168, April 1997.

[30] Woods-Vedeler, J. A, Pappa, R. S., Jones, T. W., Spellman, R., Scott, W., Mockensturm, E. M, Liddle, D., Oshel, E., and Snyder, M., "Concept Definition Study for In-Space Structural Characterization of a Lightweight Solar Array: Final Report," NASA Technical Memorandum, NASA-TM-2002-211629, Mar. 2002.

[31] Pappa, R. S., Lassiter, J. O., and Ross, B. P., "Structural Dynamics Experimental Activities in Ultralightweight and Inflatable Space Structures," *J. Spacecraft and Rockets*, Vol. 40, No. 1, Jan./Feb. 2003, pp. 15–23.

[32] Derbès, B., "Case Studies in Inflatable Rigidizable Structural Concepts for Space Power," 37th AIAA Aerospace Sciences Meeting, Reno, NV, AIAA Paper 1999-1089, Jan. 1999.

[33] Watson, J. J., "Static-Test Results for the Characterization of Inflatable Rigidizable Columns," 42nd AIAA/ASME/ASCE/AHS/ASC Structures, Structural Dynamics and Materials Conference, Seattle, WA, AIAA Paper 2001-1269, April 2001.

[34] "NGST 1/2-Scale Inflatable Sunshield Deployment," Final Report to NASA, ILC Dover Inc. and L'Garde Inc., June 1998.

[35] Lienard, S., Johnston, J. D., Ross, B., and Smith, J., "Dynamic Testing of a Subscale Sunshield for the Next Generation Space Telescope (NGST)," 42nd AIAA/ASME/ASCE/AHS/ASC Structures, Structural Dynamics, and Materials Conference, 2nd Gossamer Spacecraft Forum, Seattle, WA, AIAA Paper 2001-1268, April 2001.

[36] Slade, K., Belvin, K., and Behun, V., "Solar Sail Loads, Dynamics, and Membrane Studies," 43rd AIAA/ASME/ASCE/AHS/ASC Structures, Structural Dynamics, and Materials Conference, Denver, CO, AIAA Paper 2002-1265, April 2002.

[37] Slade, K., Belvin, W., Tetlow, T., and Behun, V., "Dynamic Characterization of a Subscale Solar Sail Using Non-Contacting Excitation and Sensing," 44th AIAA/ASME/ASCE/AHS/ASC Structures, Structural Dynamics, and Materials Conference, Norfolk, VA, AIAA Paper 2003-1744, April 2003.

[38] Hossain, A., Woo, K., and Jenkins, C. H., "Static and Dynamic Analysis of Systematically Creased Membranes," *J. Advanced Materials*, Vol. 42, No. 4, 2010, pp. 22–36.

[39] Hossain, A., Jenkins, C. H., Woo, K., and Igawa, H., "Transverse Vibration Analysis of Partly Wrinkled Membranes," *J. Spacecraft and Rockets*, Vol. 42, No. 3, 2006, pp. 626–637, DOI: 10.2514/1.11327.

[40] Wilkie, W. K. et al., "Low-Cost Piezocomposite Actuator for Structural Control Applications," 7th SPIE International Symposium on Smart Structures and Materials, Newport Beach, CA, March 2000.

[41] McGowan, A. R. et al., "Aeroservoelastic and Structural Dynamics Research on Smart Structures Conducted at NASA Langley Research Center," 5th SPIE International Symposium on Smart Structures and Materials, San Diego, CA, March 1998.

[42] Pappa, R. S., Giersch, L. R., and Quagliaroli, J. M., "Photogrammetry of a 5 m Inflatable Space Antenna With Consumer-Grade Digital Cameras," *Experimental Techniques, Society for Experimental Mechanics*, Vol. 25, No. 4, July/Aug. 2001, pp. 21–29.

[43] Lassiter, J. O., and Engberg, R. C., "5 Meter Inflatable Concentrator Modal Survey Test Report," NASA Marshall Space Flight Center Report SOTV-DEV-00-069, Oct. 2000.

[44] Jenkins, C. M. H, ed., *Gossamer Spacecraft: Membrane and Inflatable Structures Technology for Space Applications, Progress in Astronautics and Aeronautics*, Vol. 191, AIAA, Reston, VA, ISBN 1563474034, 2001.

[45] Engberg, R. C., and Lassiter, J. O., "Dynamic Testing of Inflatable Spacecraft Structures," *Sound and Vibration*, June 1999, pp. 16–20.

[46] Engberg, R. C., Lassiter, J. O., and McGee, J. K., "Modal Survey Test of the SOTV 2×3 Meter Off-Axis Inflatable Concentrator," 41st AIAA/ASME/ASCE/AHS/ASC Structures, Structural Dynamics, and Materials Conference, Atlanta, GA, AIAA Paper 2000-1639, April 2000.

[47] Slade, K. N., Tinker, M. L., Lassiter, J. O., and Engberg, R., "Comparison of Dynamic Characteristics for an Inflatable Solar Concentrator in Atmospheric and Thermal Vacuum Conditions," 41st AIAA/ASME/ASCE/AHS/ASC Structures, Structural Dynamics, and Materials Conference, Atlanta, GA, AIAA Paper 2000-1641, April 2000.

[48] Slade, K., Virgin, L., and Tinker, M., "Mode Splitting in an Inflated Polyimide Cylinder with Circumferential Asymmetry," 42nd AIAA/ASME/ASCE/AHS/ASC Structures, Structural Dynamics, and Materials Conference, Seattle WA, AIAA Paper 2001-1411, April 2001.

[49] Griffith, D. T., and Main, J. A., "Modal Testing of an Inflated Thin Film Polyimide Torus Structure," Proceedings of IMAC XVIII, Feb. 2000, pp. 1035–1041.

[50] Griffith, D. T., and Main, J. A., "Experimental Modal Analysis and Damping Estimation of an Inflated Thin-Film Torus," *J. Guidance, Control, and Dynamics*, Vol. 25, No. 4, 2002, pp. 609–617.

[51] Ruggiero, E. J., "Active Dynamic Analysis and Vibration Control of Gossamer Structures Using Smart Materials," Master's Thesis, Virginia Polytechnic Institute and State University, etd-05072002-180854, May 2002.

[52] Park, G., Ruggiero, E., and Inman, D. J., "Dynamic Testing of Inflatable Structures Using Smart Materials," *Smart Materials and Structures*, Vol. 11, No. 1, 2002, pp. 147–155.

[53] Ruggiero, E. J., Bonnema, G. T., and Inman, D. J., "Application of SISO and MIMO Modal Analysis Techniques on a Membrane Mirror Satellite." Proceedings of 2003 ASME International Mechanical Engineering Congress and Exposition, Aerospace Division, Washington, D.C., Nov. 2003, pp. 63–69.

[54] Sodano, H. A., Park, G., and Inman, D. J., "An Investigation into the Performance of Macro-Fiber Composites for Sensing and Structural Vibration Applications," *Mechanical Systems and Signal Processing*, Vol. 18, No. 3, 2003, pp. 683–697.

[55] Ruggiero, E. J., Park, G., and Inman, D. J., "Multi-Input Multi-Output Vibration Testing of an Inflatable Torus," *Mechanical Systems and Signal Processing*, Vol. 18, No. 5, Sept. 2004, pp. 1187–1201.

[56] Song, H., Smith, S. W., and Main, J. A., "Dynamic Testing of an Inflatable, Self-Supporting, Unpressurized Thin Film Torus," *J. Guidance, Control, and Dynamics*, Vol. 29, No. 4, July/Aug. 2006, pp. 839–845.

[57] Black, J. T., Smith, S. W., Leifer, J., and Bradford, L. J., "Measuring and Modeling the Dynamics of Stiffened Thin-Film Polyimide Panels," *J. Guidance, Control, and Dynamics*, Vol. 31, No. 3, May–June 2008, pp. 490–500, DOI: 10.2514/1.32236.

[58] Adetona, O., Horta, L. G., Taleghani, B. K., Blandino, J. R., and Woods, K. J., "Vibration Studies of an Inflatable/Rigidizable Hexapod Structure with a Tensioned Membrane," 44th AIAA/ASME/ASCE/AHS Structures, Structural Dynamics, and Materials Conference, Norfolk, VA, AIAA Paper 2003-1737, April 2003.

[59] Leifer, J., Black, J. T., Smith, S. W., Ma, N., and Lumpp, J. K., "Measurement of In-Plane Motion of Thin-Film Structures Using Videogrammetry," *J. Spacecraft and Rockets*, Vol. 44, No. 6, Nov.–Dec. 2007, pp. 1317–1325, DOI: 10.2514/1.25566.

[60] Murphy, D. M., "Validation of a Scalable Solar Sailcraft System," *J. Spacecraft and Rockets*, Vol. 44, No. 4, July–Aug. 2007, DOI: 10.2514/1.23024.

[61] Schuler, Charli, "Flying with Nature's Own Fuel," *NASA Jet Propulsion Laboratory*, 24 May 2003, https://www.nasa.gov/vision/earth/technologies/solarsails.html [retrieved April 2017].

[62] Murphy, D., McEachen, M., Macy, B., and Gaspar, J., "Demonstration of a 20-m Solar Sail System," 46th AIAA/ASME/ASCE/AHS/ASC Structures, Structural Dynamics and Materials Conference, Structures, Structural Dynamics, and Materials, AIAA-2005-2126, DOI: 10.2514/6.2005-2126.

[63] Gaspar, J. L., Behun, V., Mann, T., and Macy, B., "Testing of a 20-meter Solar Sail System," NASA Technical Report 20050245054, 2005.

[64] Heaton, A. F., Braafladt, A., and Artusio-Glimpse, A.B., "Validation of Solar Sail Simulations for the NASA Solar Sail Demonstration Project," AIAA SPACE 2014 Conference and Exposition, AIAA 2014-4207, DOI: 10.2514/6.2014-4207.

[65] Taleghani, B., Lively, P., Gaspar, J., Murphy, D., and Trautt, T., "Dynamic and Static Shape Test/analysis Correlation of a 10 Meter Quadrant Solar Sail," 46th AIAA/ASME/ASCE/AHS/ASC Structures, Structural Dynamics and Materials Conference, Structures, Structural Dynamics, and Materials, AIAA-2005-2123, DOI: 10.2514/6.2005-2123.

[66] Tyson, N., *Space Chronicles, Facing the Ultimate Frontier*, W. W. Norton and Company, New York, 2012, p. 165.

[67] Murphy, D., "The SCARLET Solar Array: Technology Validation and Flight Results," *Deep Space 1 Technology Validation Symposium*, Pasadena, CA, Feb. 2000.

[68] Rayman, M. et al., Dawn: "A Mission in Development for Exploration of Main Belt Asteroids Vesta and Ceres," International Astronautical Congress, IAC-04-Q.5.05, Vancouver, Canada, Oct. 2004.

[69] Vision and Voyages for Planetary Science in the Decade 2013–2022, Committee on the Planetary Science Decadal Survey, Space Studies Board, ISBN-10: 0-309-20954-4, March 2011.

[70] Murphy, D., Eskenazi, M. I., McEachen, M. E., and Spink, J. W., "UltraFlex and MegaFlex—Advancements in Highly-Scalable Power," 3rd AIAA Spacecraft Structures Conference, AIAA SciTech Forum, AIAA-2016-1947, DOI: 10.2514/6.2016-1947.

[71] Murphy, D., "MegaFlex—The Scaling Potential of UltraFlex," 53rd AIAA/ASME/ASCE/ AHS/ASC Structures, Structural Dynamics & Materials Conference, 13th Gossamer Spacecraft Forum, AIAA-2012-1581, April 2012, DOI: 10.2514/6.2012-1581.

[72] Proposal Information Package for the In-Space Propulsion Technologies, Cycle 2 ROSS 2002 NRA Amendment, http://www.spacetransportation.com/code_s/ISPT_PIP2003final.pdf, Sept. 2002, p. 14.

[73] Jenkins, C. H., Kalanovic, V. D., Padmanabhan, K., and Faisal, S. M., "Intelligent Shape Control for Precision Membrane Antennae and Reflectors in Space," *Smart Materials and Structures*, Vol. 8, 1999, pp. 1–11.

[74] Jenkins, C. H., and Tampi, M., "Local Membrane Vibrations and Inflatable Space Structures," *Space 2000*, 2000, DOI: 10.1061/40479(204)50.

[75] Price, H., "Solar Sail Technology NMP ST7 Opportunity," http://nmp.jpl.nasa.gov/st7/ST7_TA_Solar_Sail.pdf, May 2001, p. 13.

[76] Jenkins, C. H. M., Gough, A. R., Pappa, R. S., Carroll, J., Blandino, J. R., Miles, J. J., and Rakoczy, J., "Design Considerations for an Integrated Solar Sail Diagnostics System," 45th AIAA/ASME/ASCE/AHS/ASC Structures, Structural Dynamics & Materials Conference, AIAA-2004-1510, DOI: 10.2514/6.2004-1510.

CHAPTER 7

Space Flight Testing

Jeremy A. Banik
Air Force Research Laboratory, Kirtland AFB, NM 87117

Benjamin J. Urioste
Applied Technology Associates, Albuquerque, NM 87123

7.1 INTRODUCTION

Previous chapters have described the best practices of testing ultra-lightweight spacecraft structures in the terrestrial environment. Technologists go to great lengths to simulate the space environment with gravity off-load fixtures, thermal-vacuum chambers, radiation exposure facilities, dynamic load actuators, and non-contact metrology. Each of these have limitations when implemented individually, much more so if ever implemented simultaneously. The conclusion is that spaceflight is the only way to test in a truly combined representative environment. This chapter discusses the unique benefits, risks, and best practices for testing large ultra-lightweight structures in space.

Since 1990, 22 major space flight tests have been conducted involving ultra-lightweight spacecraft, as documented in the open literature. These cases form the basis upon which the following discussions are developed. At the end of the chapter, a series of tables documents the plans and approaches used for each test case, including the following details: mission objectives, test article, excitation method, sensing approach, launch and orbital operations, and post-mission data analysis. Further reading on each is available in the references.

7.2 PREPARATION

INTRODUCTION

Leading a technology through a space flight test is exciting and can be deeply rewarding, but not every new large space structure technology *should* be flow in space. It is one of the most expensive tests an engineer will undertake so those considering to advocate for this testing may benefit by first asking themselves the following: 1) "Why would someone want to fly my new technology?" Once answered, the next question to ask is, 2) "What do I need to learn in order to convince myself or the customer that this technology is worthwhile?", and finally

3) "How do I find a launch?" The following short sections should help the reader to navigate answers to these questions.

7.2.1 DEFINING THE NEED

Deployable structures have been notoriously high risk to spacecraft mission success. Many reflector and solar array deployment anomalies have been documented in the open literature despite our decades of heritage in articulated joints, spring and motor actuation methods, damping approaches, and mesh reflector systems. Studies have shown that 32 percent of space mission degradation has been due to mechanism and thermal causes, second only to electrical causes, which account for 45 percent of degradation [1]. Some recent commercial examples of deployment anomalies are as follows:

- "For SES-4, the problem appears to have been that one of the two side panels on one of the two five-panel solar arrays did not lock into place immediately on deployment. Deployment eventually occurred, with the panels locking into place [2]."
- "Loral-built Estrela do Sul-2 (EDS-2) satellite launched in May 2011 failed to deploy fully one of its two arrays, reducing the satellite's power and communications capacity. Insurers have since paid a $132.7 million claim [2]."
- The SkyTerra-1 22-m-diameter antenna had deployment challenges that were resolved within 10 days [3].

These are just a few of the public examples. There are many more private examples.

The natural reaction to such anomalies is "risk analysis paralysis," an extreme resistance to trying something new, as suggested by Kennedy [4]. He proposes that a partial solution—albeit a cultural one—to this paralysis is to accept loss as normal business: "Aversion to failure or loss has become so endemic to our space enterprise that programs are often very late, deeply over budget, or canceled. Acknowledging that the possibility of loss or failure is part of the space equation is the only way to break this cycle."

Of course, this extreme aversion to risk in space is a complex issue, warranted in some cases when failure has grave consequences, yet in other cases this aversion seems a product of strong budgetary and political forces beyond the control of satellite technology developers. In any case, the technology developers maintain the responsibility to prevent structures and mechanism anomalies from degrading missions.

We often prevent anomalies only through testing. Sometimes ground-based testing is adequate; other times space-based testing is required. In either case, the engineer must scale the test expense proportional to the risk posture of mission managers and customers.

Previous chapters in this book have pointed to the impossible reality of simulating the fully combined environment of space on the ground, most significantly the thermal, vacuum, and micro-gravity. Space flight testing remains the only way to uncover the unique problems that arise from these combined effects.

Two cases illustrate this point. The first is the discovery that mechanically hinged structures tend to be more highly damped in micro-gravity than on the ground due to the non-linear free-play effects of joint gaps, a lesson learned on the Middeck 0-gravity Dynamics Experiment (MODE) [5]. Secondly, the Inflatable Antenna Experiment (IAE) taught us, among other things, that 1) small, second-order internal forces can have a major impact on the deployment of large lightweight structures and 2) perturbations and damping inherent in gravity off-loaders may swamp internal structural forces and hide the resultant effects [6] on deployment success or the lack thereof. The following sections describe other space flight tests that have revealed critical structural behaviors previously masked during ground testing. A deeper look into these cases reveals many practical guidelines for those who are planning the next space flight test.

7.2.2 DEVELOPING A PLAN

One of the main reasons to test in space is to discover the "unknown unknowns." Planning for such a test requires anticipating these unexpected behaviors. If anticipated, these events can be quantified *in situ* with carefully placed sensors. Answering the following questions may help to define the purpose of, and thus the measurements needed for, such a space flight experiment:

1. *What do we expect to learn from a space flight test that we cannot learn on the ground?* This question aims to define the structure's critical functions that are not well understood from ground testing yet must be understood before flying on a demonstration or operational mission. Listing the highest technical risks is a good place to start.
2. *What conclusions do we expect to draw from a successful flight?* While the answer to the first question is broad, the answer to this question is much more specific because it should include quantifiable success criteria. Flowing out of that criteria is a list of flight objectives. Using verbs such as "test," "measure," "capture," and "collect" to define flight objectives can be helpful. Avoid the use of "demonstrate" unless the goal of the mission is indeed a demonstration (as defined shortly). Example objectives are provided in the case study tables near the end of this chapter.
3. *What will we do with the flight data?* The purpose of this question is to think through how the flight data will be analyzed and compared to ground testing and analysis. Sometimes, test results will not be ready in time to incorporate any lessons learned into the next unit. From a practical perspective, one might then ask, "Why do the test?"
4. *What on-orbit measurements do we need in order to meet objectives?* Answering this question requires some thought around sensor placement, quantity, sampling rate, noise floor, and actuation method. Space environment thermal, radiation, and atomic oxygen conditions are also considered here.

For example, is active temperature control or passive solar pointing required, or, perhaps, is only temperature knowledge required? What will my structure require of the host spacecraft in terms of power draw, thermal control, data volume, data processing, pointing precision, pointing knowledge, or pointing torque?

5. *What can I avoid or adopt from history?* This chapter was, in fact, born from a need to answer this question. Although it is seemingly obvious, historical experiments can teach us to avoid mistakes and adopt best practices. The value of historical study has been echoed by many great thinkers; the following sentiment is certainly not unique:

> "Those that fail to learn from history are doomed to repeat it."
> —Winston Churchill

Defining terms may seem an onerous and unnecessary exercise at times. However, the distinction between an experiment, a test, and a demonstration, within the context of space flight, is central to expectation management across all stakeholders—from the engineers to the funding agents. For example, past space flights (such as the IAE) that were expected to operate as demonstrations have generally had lower success rates than those planned and operated as space flight tests or experiments. Success is defined based on the expected outcome which should be agreed upon long before launch. Originally defined in Chapter 1 of this book, here is a review of these three terms:

- *Experiments* are methods of pure research. Experiments are done to help understand what is not known, say, of some physical phenomenon. Of the three activities, the outcomes of experiments are the least deterministic: "In the case of an experiment, one may not even know what it means for it to work—the experiment has the capacity to surprise us." Given this high level of uncertainty, the experiment remains the most flexible procedurally. In the annals of science, experiment has led the way in developing theories that explain the world around us.
- Like experiments, *tests* are also part of research, but they have more of an applied bent. Tests are formalistic procedures that produce data; the test formalism is often codified in standards. Thus, data from tests carry a higher degree of confidence than data from experiments, leading to increased results utilization.
- A *demonstration* is "an event orchestrated to show how a process or product works." "Demonstrations are designed to educate and convince once the exploration has been done and the discoveries have been made. Demonstrations have the power to convince because of the smoothness of performance, distancing the audience from the untidy craft of the scientist."

As these definitions suggest, the term *demonstration* should be used most sparingly within the context of this chapter. A flight demonstration (whether explicitly defined or functionally operated as such) is not designed to capture unexpected structural behaviors. When flight goals are not well defined or controlled, flight results are confusing, providing little useful data for direct application to future designs. Or when unexpected events occur, there is insufficient instrumentation to isolate and identify the phenomena in order to prevent similar events on future missions. Wada concluded the following from his review of flight experiments [5].

> Many of the [historical structures-related] space experiments did not significantly contribute to the engineering knowledge to help validate predictive models. The objectives of the ambitious [historical structures-related] experiments were more a demonstration rather than to obtain knowledge valuable for future designs. Often the most valuable experiments were those that focused on measuring single phenomenon that could not be measured on the ground. The experiments were simplified to isolate the desired measurements from other uncontrollable disturbances or phenomena.
>
> Often when the experimental goals are not well defined or controlled during an experiment, the results are confusing and results in little useful data for direct application to future designs. When unexpected events occur, often insufficient instrumentation exists to isolate and identify the phenomena to prevent similar events on future missions. The use of space experiments to validate operational system prior to flight is generally technically unrealistic and does not meet the cost and schedule constraints. However, if properly designed, [space experiments] are very important to help validate the predictive tools related to structural performance and its interaction with the environment, control and navigation system.

An historical review of ultra-lightweight space structures experiments and tests teaches us to keep it simple. Trying to do too much with too little instrumentation leads to confusion in tracing the root cause of an anomaly, a common pitfall. In summary, here are some guidelines to follow:
Avoid these:

- General goals that focus on mission demonstration
- Insufficient instrumentation that captures only the ideal and expected behavior
- Measurements that cannot distinguish between uncontrolled external disturbances and the controlled inputs

Pursue these:

- Specific goals that focus on model validation
- Abundant instrumentation that captures both the expected and the unexpected
- Focused measurements that isolate the controlled inputs from the uncontrolled

7.2.3 MANIFESTING A LAUNCH

One of the biggest challenges in achieving space flight is securing a launch to orbit. Yet recent commercial and government interest in satellites in the 10 kg to 300 kg class has grown sharply in the last decade, and this interest has improved space access for smaller payloads. As the number of legitimized mission sets have grown so has the low-cost launcher market. Competitors in both ride-share services and dedicated low-cost launch have emerged. For example, Spaceflight Industries provides an international launch brokerage service, openly advertising firm-fixed-price launch slots. At the time of this publication, $995,000 buys a launch to Low Earth Orbit (LEO) for a 12-liter, 20-kg containerized bus [7]. New dedicated launchers are emerging into the markets as well. Virgin Galactic promises to soon offer responsive dedicated launch of a 300-kg spacecraft to LEO for $18 million [8]. Other launch providers promise similar price points, as identified by the Tauri Group's annual assessment of the international space launch market [9].

Assuming your space flight experiment is ready for launch, buying access to space may still be cost prohibitive. Furthermore, you may not have access to a ground command and control network needed to operate your experiment. In this case, both NASA and the U.S. Department of Defense (DoD) provide a means to pay for integration, launch, and operations costs, provided the experiment aligns with the respective agency's goals.

For example, the DoD Space Test Program (STP) is charged with providing access to space for experiments on the DoD Space Experiments Review Board (SERB) list. Additionally, STP is the "front door" for all non-DoD Auxiliary Payloads seeking launch opportunities on DoD missions. In this capacity, STP manages the launch integration of a wide variety of missions, including DoD satellites and those built, owned, and operated by NASA, civilian universities, commercial entities, and foreign governments. Often, a single launch mission has multiple satellites from multiple organizations and may contain payloads from one organization hosted on satellites owned by another [10].

The NASA Educational Launch of Nanosatellites (ELaNa) is similar in goal to the DoD STP program but is limited to CubeSat form factors. ELaNa provides opportunities for small auxiliary payloads to fly as ride-shares on previously planned missions [11].

Various flight platforms are used to conduct space flight tests. The most common approach for a large lightweight structure, as documented in the case studies to follow, is a dedicated free-flyer such as the large deployable reflector (LDR). However, the International Space Station (ISS) also provides platforms for internal and external (deployable) experiments. Externally, the ISS provides a CubeSat launcher, racks for long-duration experiments, and more recently the Roll Out Solar Array (ROSA) experiment unfurled from the end of the robotic manipulator, to be detailed in the next section.

7.3 CASE STUDIES

This section presents a detailed assessment of four space flight test cases followed by an expanded tabulation of an additional 18 cases. The first four detailed cases were selected based on adequate literature available to document their approach and the variety of structural precision. The first two deal with moderate precision antenna systems: the 19-m Japanese LDR and the 60-m U.S. Shuttle Radar Topography Mission (SRTM), both of the year 2000. The second two cases are low-precision solar array examples: the Solar Array Flight Experiment (SAFE) of 1984 and the ROSA of 2017.

7.3.1 LDR

The LDR on the Engineering Test Satellite (ETS-VIII) was launched in December 2006 on an H-IIA vehicle [12, 13]. Born out of a family of development reflector systems, LDR benefited from a methodical series of space flight tests.

LDR is constructed from hexagonal self-supporting modules, apparent in Fig. 7.1, that may be independently assembled and adjusted for surface precision. The aperture surface is a gold-plated molybdenum mesh covered by a structural cable surface with adjustment ties back to a secondary cable network (behind the mesh). The secondary network is anchored to a radial arm deployment structure. Six radial arms each consist of an upper and lower member connected in a four-bar linkage arrangement stiffened by a diagonal tension member in the center. Deployment is accomplished by paying out a cable that allows the diagonal member to slide along the center axis via stored energy in a sliding spring mechanism.

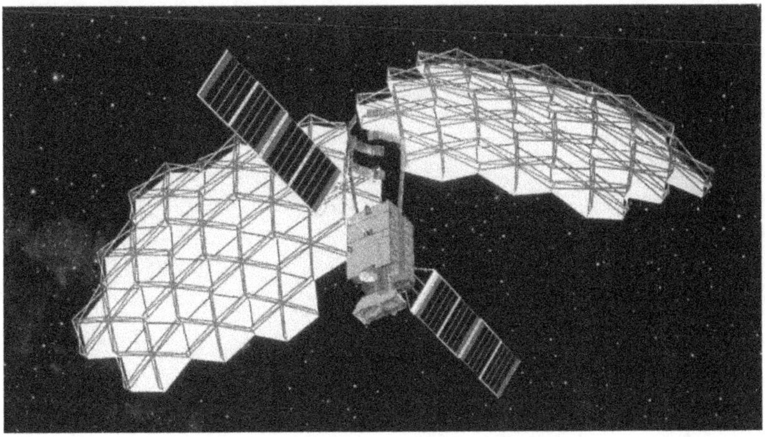

Fig. 7.1 In-orbit configuration of LDR on the ETS-VIII spacecraft. (© JAXA [Printed with permission].)

The first version of LDR was a space flight test called LDREX, a scaled version of LDR at 7 m in diameter and constructed with seven 2.5-m modules [14]. The purposes of LDREX, launched in 2000, were to observe deployment behavior and to evaluate accuracy of analytical predictions. LDREX failed to deploy when the mesh snagged on the truss members; this failure was attributed to unexpectedly large vibrations in the reflector hold-down mechanism during launch.

The second version of LDR, a space flight test known as LDREX-2, was launched in 2006. LDREX-2 was the same size as the original LDREX but with the new purpose of confirming that design changes had improved deployment reliability, and, again, with the aim of evaluating the accuracy of analytical predictions [12, 13]. The ground test deployment of LDREX-2 is depicted in Fig. 7.2. After a successful ground and then space flight deployment and model correlation effort, LDREX-2 was scaled up to the full-size LDR. Then, it was ground tested, stowed, and launched as a flight demonstration on the ETS-VIII spacecraft. Figures 7.3 and 7.4 show the ground test and stowed configurations of LDR, respectively, and Fig. 7.5 shows the on-orbit deployed reflector. At 19 m by 17 m, each of the two LDR reflectors per satellite were constructed with 14 modules that were 5 m in diameter. Stated objectives of the ETS-VIII mission were to demonstrate the following capabilities: 1) 3-ton-class satellite bus technology, 2) large deployable antenna reflector, 3) satellite audio and data communications with mobile handheld terminals, and 4) geostationary satellite positioning.

All three space flights used camera imagery to observe deployment behavior. In addition to imagery, LDREX-II and LDR included a means to measure the deployment cable tension and linkage deployment angle.

Fig. 7.2 LDREX-2 partially deployed. (© JAXA [Printed with permission].)

Fig. 7.3 LDR fully deployed. (© JAXA [Printed with permission].)

The success of LDR was due in no small part to the persistence of mission managers who were willing to fly a second unit after experiencing a major anomaly on the first. After the deployment failure on LDREX, the program persisted to fly LDREX-2, which successfully mitigated the deployment risk from

Fig. 7.4 LDR stowed for launch. (© JAXA [Printed with permission].)

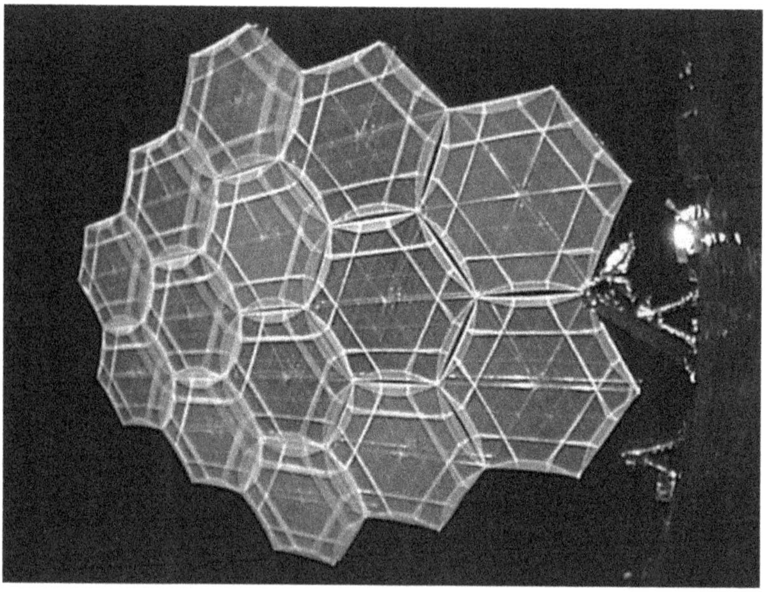

Fig. 7.5 LDR successful deployment on-orbit. (© JAXA [Printed with permission].)

LDREX. Only after the risk had been mitigated did the program move forward with a successful flight demonstration of LDR on ETS-VIII. This case is a good example of the differences between flight testing, LDREX and LDREX-2, and a flight demonstration, LDR.

7.3.2 SRTM

The SRTM was launched in February of 2000 on the Space Transportation System Mission 99 (STS-99). Although primarily focused on radio frequency Earth sensing, this mission also tested the deployment, stability, and dynamics of an extremely stiff mast. The 60-m-long (nearly 200-ft-long) Able Deployable Articulated Mast (ADAM) is shown in Fig. 7.6. Developed by AEC Able, it uses four graphite longerons and an articulated deployment scheme similar to the Folding Articulated Square Truss (FAST) that currently supports the solar array wings on the ISS, though the FAST mast uses aluminum longerons and is 33 m long.

The two objectives of the SRTM mission were to 1) successfully deploy the 60-m-long mast and demonstrate functionality of the mast vibration damping system and 2) survey a topographic database of 80 percent of the Earth's land surface.

To achieve these objectives, the ADAM mast deployed an outboard antenna 60 m from the Space Shuttle Orbiter payload bay, shown in Fig. 7.7, creating a stable, fixed separation from a second antenna inboard the payload bay. Together,

Fig. 7.6 ADAM mast. (Image Credit: NASA JPL.)

these antennas form a single-pass interferometric synthetic aperture radar, where the inboard antenna is used in transmit and receive mode (whereas the outboard is receive-only). The antennas are based on a previous experiment, flown in April 1994 on STS-59 and known as the Spaceborne Imaging Radar-C/X-band Synthetic Aperture Radar (SIR-C/X-SAR).

Stability requirements on the mast were stringent. The tip misalignment was required to be less than 0.3 deg in pitch and yaw. A 2 degree of freedom (DOF) actuator at the tip was used to actively compensate for quasi-static misalignment. To measure that misalignment, a high-precision star tracker was mounted near the inboard antenna to track three LED targets mounted on the outboard antenna. The red LED targets were mounted on hollow graphite-epoxy stalks separated from each other by 1 m. The star tracker camera, which generated LED

Fig. 7.7 SRTM from the Orbiter payload bay. (Image Credit: NASA JPL.)

centroid information at a 4 Hz sampling rate, provided real-time relative attitude and position between the outboard and inboard antennas. In addition to antenna pattern compensation, this metrology system was used to measure the mast deployed position and the mast structural dynamics response [15].

Exciting the mast structural dynamics was done using the Orbiter Vernier attitude control system (VCS) thrusters and the Orbiter reaction control system. The Orbiter attitude sensors and the star tracker measurements of the mast were used to solve for the modal frequencies using power spectral density functions. Knowledge of the mast frequencies in-flight needed to be accurate within 5 percent to minimize attitude control maneuvers of the orbiter.

Accurately predicting the natural frequencies of structural vibration is important for most large ultra-lightweight space structures but was particularly important for this mission because errors meant wasted Orbiter fuel. Due to the practical difficulties of performing precision modal tests on the integrated mast system in a terrestrial gravity environment, the pre-flight estimates of modal frequencies are accurate only to within 10 to 20 percent. Notch filter settings in the Orbiter's attitude control system are selected to reduce mast response, and any uncertainties in pre-flight estimates lead to inefficient control maneuvers, which ultimately wastes fuel.

SRTM successfully used the method for in-flight modal tests that was previously used in 1984 on SAFE, combining the host vehicle attitude determination system with payload optical-based relative motion tracking. These tests were necessary to improve our knowledge of the ADAM mast structural frequencies and to allow for near-real-time attitude control system optimization. In general, errors in our ability to predict structural frequencies and mode shapes leads to over-sized attitude control systems or, in some cases, the loss of vehicle control.

7.3.3 SAFE

The SAFE was launched in August 1984 on STS-41D. The specific purpose of the SAFE test, one of multiple experiments in the OAST-1 payload, was to characterize the deployment, retraction, and deployed dynamics of a flex-blanket solar array supported by a single column. As a precursor to the current flex-blanket solar array wings used on the ISS, SAFE deployed and tensioned a similar sized, accordion folded blanket: 31.5 m × 4.0 m. The mast constructed by AEC Able Engineering Inc., used three coiled glass-fiber longerons, in contrast to the articulated aluminum longerons used in the FAST mast columns that currently support the ISS wings.

In the short 15-hour mission, this structure was deployed as shown in Fig. 7.8 then subjected to structural dynamics experiments in both orbital day and night. The wing in-plane and out-of-plane motions were excited with the Orbiter Vernier attitude control system (VCS) thrusters similar to the approach later used on SRTM. In addition, the response was measured with micro-g servo accelerometers in the tip box and 23 retroreflector targets on the blanket. These targets

Fig. 7.8 SAFE partially deployed from the Orbiter cargo bay. (Image Credit: NASA Archive.)

varied in diameter from 0.55 in. to 1.6 in., and the targets were spaced proportional to the distance from the root, longer spacing at the tip, shorter near the root. Data were captured and stored using a data acquisition system (DAS) with a tape recorder.

For the daytime experiments, images were captured with video cameras. These images were post-processed using photogrammetry techniques to extract the frequencies and mode shapes of natural vibration as well as the wing shape during the natural temperature changes of an orbit.

During orbital night, five laser diodes illuminated 18 of the retroreflector targets. The 30 mW diodes broadcast at a wavelength of 820 nm. A 256 × 256-pixel solid state charge injection device captured the reflections. This nighttime mode of the OAST-1 experiment was known as the Solar Array Flight Dynamics Experiment (SAFDE) [16].

The on-orbit deployed fundamental frequency was measured to be in the range of 0.059 Hz to 0.072 Hz depending on blanket tension variability caused by wing orientation to full-sun or eclipse. During night, an unexpected blanket curvature and mast twist was observed, reducing the correlation accuracy of data compared to ground predictions. The mode shapes of natural structural vibration were observed to be very similar to a clamped-free beam since there were no noticeable torsion or transverse bending components to these blanket mode shapes. Only out-of-plane sinusoidal shapes were observed. Estimated displacement error in the photogrammetry analysis was less than 0.1 in. at a distance of 100 ft from source to target. The first-mode critical damping was measured to be 3.5 percent, which was much higher than the expected 0.5 percent. This observation is consistent with previous observations from mechanically articulated structures in micro-gravity. For example, the MODE experiment showed higher damping versus terrestrial due to the non-linear effects of free-play in mechanical joints and the friction-induced hysteresis during that free-play motion [5].

SAFE was the first large-scale space experiment to use photogrammetry to capture structual dynamics during the day and laser diodes at night. As such, SAFE paved the way for others to use this technique. A key finding for similar tension-based architectures was the differences in structural frequency due to temperature change and thus tension of the blanket. Chapter 6 also discusses the SAFE and OAST-1 experiments.

7.3.4 ROSA

Roll out Solar Array (ROSA) is an external experiment on the International Space Station that was launched June 3, 2017 on the 11th NASA commercial resupply mission (CRS-11); the experiment was then operated from June 17 to 25, 2017. The goal of the ROSA experiment was to unfurl this new solar array system for the first time on-orbit then measure the structural dynamics behavior and power production performance in the combined microgravity and extreme thermal environment of space [17, 18].

Figure 7.9 shows the 5.36 m long by 1.67 m wide ROSA wing deployed. It is supported by two reverse-rolled, 4-in.-diameter, carbon composite slit-tube booms, more traditionally known as storable tubular extendible members (STEMs). The booms are comprised of a three-ply carbon fiber reinforced epoxy laminate. The booms were pre-flight tested to thermal cycles from 90°C to −65°C while measuring startup torque to verify positive margins despite the stress-relaxation expected while stowed prior to and during launch and transit to station. Attached to the root tube and tip mandrel, the blanket is constructed of an open weave fiber mesh populated with aluminum mass simulators in place of photovoltaic cells except for six electrically active triple junction crystalline SPMs attached at the center, root, and distal end of the blanket. Deployment rate is controlled by dampers integrated in the tip mandrel. Nineteen pounds of blanket tension is maintained with constant force spring motors that allow the blanket to grow and contract during thermal excursions.

Fourteen days after launch, the wing was extracted from the unpressurized Dragon trunk then positioned into the science orientation attached to the end of the special purpose dexterous manipulator (SPDM) and moved to a predetermined orientation conducive to emergency jettison, as shown in Fig. 7.10. Up to 114 watts at 120 VDC was provided along with data communications and a dedicated video downlink to the experiment through the robotic arm tool changeout mechanisms (OTCM) at the SPDM. Once the wing was positioned, the experiment was limited to seven days of ISS-supported operations before

Fig. 7.9 ROSA flight wing fully deployed during environmental ground testing. (Image Credit: DSS, Inc.)

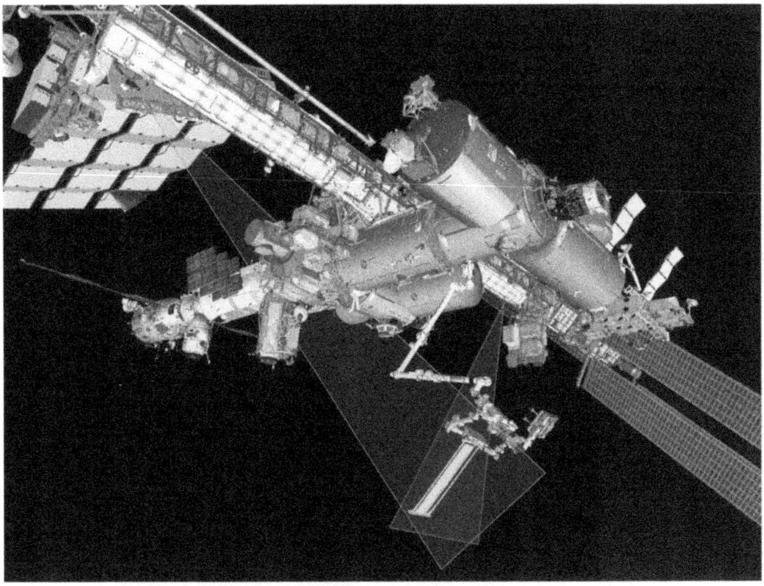

Fig. 7.10 Simulated view of ROSA with two station cameras imaging the wing while it is positioned at the end of the SPDM manipulator. (Image Credit: NASA JSC IS&G Group.)

being required to stow for disposal. This was the first unfurlable payload on the ISS robotic arm.

On Sunday, June 18, ROSA unfurled as planned in 3 minutes 34 seconds under the power of the stored strain energy in the high strain composite slit tube booms. Deployment was nominal, indicating the mechanisms and composite booms were healthy after 10 months of storage, launch, and 215 on-orbit thermal cycles. Figure 7.11 shows the deployed wing as recorded by one of five ISS cameras. Although not pictured, four on-board wide field of view (FOV) cameras were mounted at the ROSA base structure to record deployment first motion, dynamics excitation oscillations, and retraction latch engagement.

Over the next four days after deployment ROSA completed 104 forced excitation dynamics characterization runs. The wing base was oscillated with a motor and linear ball screw actuator. A series of pre-programmed sinusoidal sweeps excited the wing fundamental bending and blanket drum frequencies during full solar illumination and again during solar night. Nine accelerometers and 21 resistance temperature detectors (RTD) captured the response of the wing in real time at a reduced sampling rate of 20 Hz due to data bandwidth limitations of the SPDM bus. Full-rate 200 Hz accelerometer data was downlinked during off-hours. The blanket was populated with 84 flat 1-in-diameter matte white acrylic decals for photogrammetry; the booms were painted with 20 1-in. targets using a silver metallic Sharpie®; the spreader bar on the tip mandrel had four spherical 1.75-in. blaze orange Delrin® targets. The base plate structure had approximately 20 additional white decals for additional photogrammetry reference as shown in the Fig. 7.12 image of the stowed payload prior to launch. During experiment operations, the team was able to isolate two primary

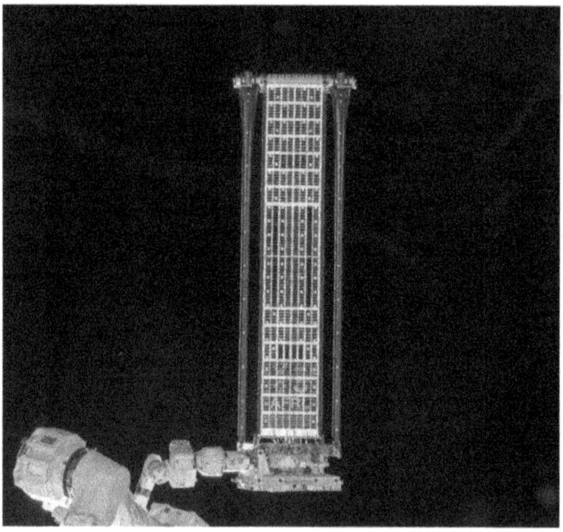

Fig. 7.11 ROSA unfurled on June 18, 2017. (Image Credit: Matt LaPointe, DSS.)

Fig. 7.12 ROSA flight payload includes the stowed wing (top), delivered by Deployable Space Systems, the base plate (middle), integrated by the DoD Space Test Program, and the Express Pallet Adapter (bottom). provided by NASA JSC. (Image Credit: DoD Space Test Program.)

modes: the primary bending mode and the blanket drum mode. A secondary blanket mode was also confirmed through camera imagery.

The team had access to five station cameras during the mission. Images captured during the dynamics events were post-processed using photogrammetry techniques to extract the three-dimensional position of all targets and thus the frequencies and mode shapes of natural vibration. The actual excitation amplitude and phase profiles were planned to be captured by a triaxial accelerometer at the wing base, but high frequency noise from the motor was too high to resolve these input profiles.

In addition to the forced excitation, the wing dynamics were monitored during the rapid heating and cooling typical of transition from orbital night-to-day and day-to-night. No thermally induced vibration response was observed, as none was expected for the thermally stable carbon composite boom structure.

Aside from deployment, capturing the structural dynamics response was the primary purpose of the ROSA flight. However, the blanket did include six active triple-junction crystalline photovoltaic SPMs from three different vendors. The launch and deployment survivability of these SPMs was assessed by periodically measuring the temperature and the voltage output of each across a shunt resistor. Knowledge of solar incidence angle with respect to the wing surface is provided within 10-deg accuracy by ISS telemetry. Achieving normal solar incidence proved difficult due to the specific wing orientation required for safe jettison; therefore, the wing was moved to another orientation for several hours to measure the SPM voltage output.

ROSA attempted retraction on June 25. Three successful retraction and re-deployment events proved the additional robustness and repeatability of the rollable booms, mechanisms, and live solar cell strings where no voltage degradation was observed when compared to previous measurements. However, a slight but repeatable telescoping misalignment prevented the wing from latching in the retracted state, a requirement for re-stowage into the Dragon trunk for disposal. Therefore the mission was concluded when NASA jettisoned the ROSA wing on June 26, the planned backup disposal operation.

In planning for the ROSA experiment solar pointing needs had to be managed. For example the station beta angle cycles from $+70^\circ$ to -75° over the course of several months. Extreme high or low beta angles cause very short eclipse durations of only a few minutes in some cases. The wing requires at least 10 minutes of soak to reach thermal equilibrium prior to a forced sine sweep excitation; therefore, ROSA required a constraint on the maximum and minimum station beta angle, restricting the experiment launch window. Additionally, shadowing on ROSA from station wings and modules is a concern because this creates an uncertain temperature and thus stress field in the blanket.

The diversity of organizations that partnered on the ROSA flight program is an example of the strong coordination required to conduct a successful space flight test of this scale.

- The Air Force Research Laboratory (AFRL) Space Vehicles Directorate provided principal experiment design, oversight, and sponsorship to ROSA.
- ROSA wing hardware design, development, construction, and testing was performed by Deployable Space Systems, Inc., while Ecliptic Enterprises Corp. developed the data acquisition and control unit. LoadPath, LLC., in conjunction with AFRL fabricated and validated the high strain composite slit tube columns and material system.
- The DoD STP Houston office brokered launch as well as funded and managed payload integration. The STP developed the jettison system and the robotic grapple and electrical interfaces. STP also led the integration of the completed payload to the ISS and to the SpaceX Dragon capsule, and they developed the command and data handling architecture needed to get from the payload to the ground and commands from the ground to the payload.
- The NASA Langley Research Center (LaRC) Advanced Measurements and Data Systems Branch as well as the Johnson Space Center (JSC) Image Science and Analysis Group provided photogrammetry metrology system design and analysis.
- The LaRC Structural Dynamics Branch assisted with structural dynamics analysis predictions and data analysis.

- Additional mission funding was provided by the U.S. Air Force Space and Missile Systems Center (SMC/AD).

A review of the ten-year ROSA development timeline in Fig. 7.13 shows that the program moved from flight manifest to post-mission data analysis in four years. But the ROSA wing was invented six years prior, highlighting the need for patience and persistence in space technology development.

Since ROSA was launched to the ISS, extra care was taken through NASA safety panel oversight to ensure the experiment would not cause harm to the station or crew. For example, a Planetary Systems Corporation Motorized Lightband separation system at the base of the wing was required in the event of a deployment problem or if the wing was not able to latch stowed for disposal, as turned out to be the case. An independent 0.2 g's accelerometer safety circuit was implemented to prevent the wing from oscillating with too much acceleration, potentially damaging the robot arm to which it is attached. Fracture critical margins had to be demonstrated for the composite STEM booms, and all externally exposed edges had to be smoothed so as not to snag during unplanned crew extravehicular activity.

During the flight experiment development and planning process, the ROSA wing was concurrently being evaluated for commercial use by Space Systems Loral (SSL) [19]. Concurrent flight qualification efforts proved beneficial to both the AFRL and SSL programs even though the detailed wing configurations were different. AFRL was able to tailor flight experiment objectives to better match the needs of the commercialization program, and since the ROSA experiment was completed by June 2017, SSL was able to use the flight test data.

7.4 COMPREHENSIVE SURVEY

The flight test approaches of the previous four and an additional 18 cases are detailed in Tables 7.1–7.5. No doubt that additional flight cases exist beyond these 22, yet these were chosen because they are available in the public domain.

Table 7.1 provides an alphabetically ordered general description of each case, including the following details: mission name, launch date, purpose, objectives, test article description, instrumentation, and the mission duration. Tables 7.2–7.5 each re-categorize the experiments according to the primary objective, listing greater detail than Table 7.1. Table 7.2 describes those experiments that measure the frequencies and mode shapes of natural vibration and lists the methodology by which that objective is achieved, including actuator type, instrumentation, sampling rate, noise, floor, and data analysis method. Table 7.3 describes methodologies for critical damping measurements. Table 7.4 describes thermally induced dynamic disturbance measurements, and Table 7.5 describes deployment force and torque measurements.

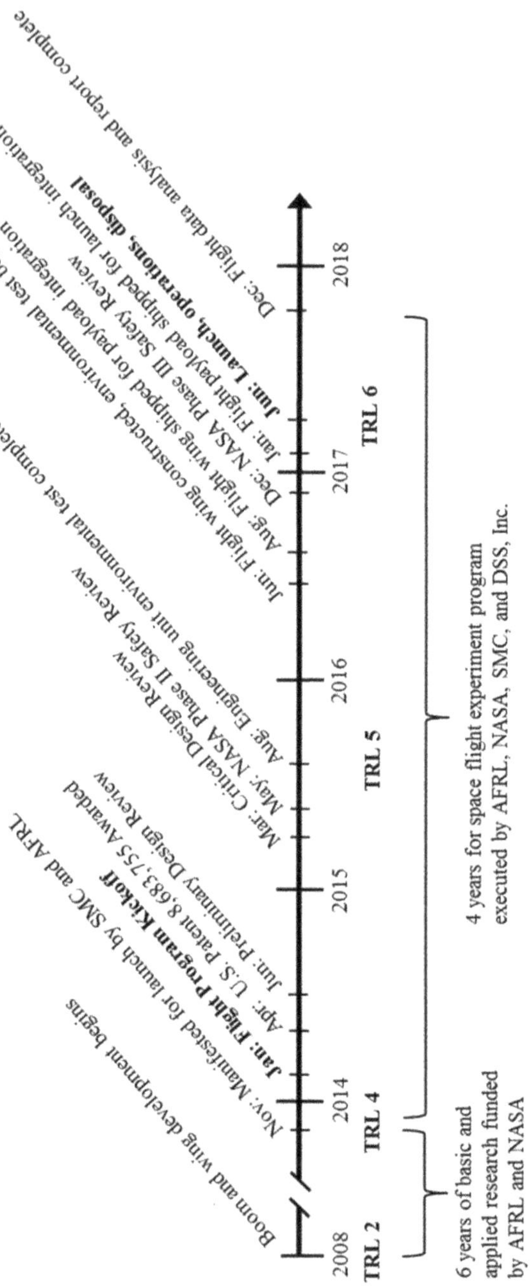

Fig. 7.13 The 10-year lifecycle of the ROSA flight experiment from concept birth to post-flight data analysis.

TABLE 7.1 OVERVIEW OF SPACE FLIGHT TEST CASES

Mission	Launch	Purpose	Objectives	Test Article	Instrumentation	Duration	Reference
Advanced Controls Technology Experiment (ACTEX I)	Classified host	Investigate the suitability of using embedded piezoceramics as sensors and actuators for structural control in the space environment.	• Demonstrate active structural control strategies. • Demonstrate system identification algorithms. • Measure structural damping. • Investigate long-duration environment survivability of the actuators in temperature extremes (−80°C to 129°C).	Variable stiffness Tripod, 10 × 12 × 22 in. with embedded piezoceramics actuators and passive visco-elastic damping materials. Control electronics designed for three-year lifetime with reliability of 0.605. Active thermal control using heaters and thermostats to maintain above − 55°C.	Piezoceramics sense and control tripod bending and torsion modes, operating in the 1–1000 Hz bandwidth, seven tip accelerometers (Endevco 7703A 1000pC/g) to recover six DOF motions of tip plate, root plate accelerometer, seven thermistors	LEO, four years, 60 on-orbit experiments	[20]
Advanced Controls Technology Experiment (ACTEX II)	DoD launch failure in June 1995 on STEP M3 spacecraft	Investigate vibration suppression using passive and active control means to reduce solar array motion and their effects on the host spacecraft in the space environment.	• Investigate controls approaches for lightly damped solar arrays wings that had to be moved quickly to avoid threats. They had a 5:1 inertia ratio in roll axis. • Operate in the combined space environment: thermal vacuum and EMI/EMC	24 × 85 in. aluminum deployable solar array mass simulator with carpenter tape hinges. The wing was joined to the base through a three-beam yoke, a two-axis gimbal, and a visco-elastically damped interface joint. Placed between the Titanium layers was a viscoelastic material by 3M: ISD 110, 112, and 113 acrylic. Each layer was carefully designed to perform at different temperature regimes: 19–120°F. The yoke is RS3/T300 cyanate ester graphite with embedded piezoceramic actuators.	Four accelerometers on the wing structure, three on spacecraft body, eight thermistors on wing	N/A	[20]

(Continued)

TABLE 7.1 OVERVIEW OF SPACE FLIGHT TEST CASES (*CONTINUED*)

Mission	Launch	Purpose	Objectives	Test Article	Instrumentation	Duration	Reference
Active Rack Isolation System (ARIS)	Space Shuttle *Atlantis* STS-79, Sept. 1996	Isolate microgravity-sensitive science experiments mounted in the International Standard Payload Rack (ISPR) from the structural vibrations present on the large space station orbital structure.	• Determine if the system would be stable when all six axes are controlled. • Determine if the system's position response to accelerometer drift would result in excessive motion. • Measure isolation performance and determine the impact of attaching umbilicals. • Measure the quiescent background acceleration levels induced by the controller itself.	A rack system that detaches ISPR from the station structure and held inertially via an active control system. Umbilicals remain connected to the rack to support power, fluid cooling, and data communication as required by science payload.	Inertial accelerometers located in three corners of the rack. Integral position sensors placed between the rack and the station. Voice coil rotary actuator used to reduce profile and power consumption.	14-minute test taken while the shuttle was docked to MIR	[21]
DragNet-1 DragNet-2	ORS-3 Nov. 2013	Demonstrate deployment of a drag de-orbit device in microgravity and after long-term stowage on-orbit.	• Test DragNet-1 deployment on a Minotaur I upper stage. • Test DragNet-2 deployment function after long-term stowage on STPSat-3 spacecraft. • Image the deployed surface using ground-based telescopes. • Test the deployment function of high-strain composite tape hinges on stand-off module.	Identical DragNet systems each comprised of four pantograph booms unfold a pyramidal-shaped polyimide sheet. Eight high-strain composite tape hinges separated the stowed DragNet module from the host structure.	No on-board instrumentation. Ground-based telescopes used to confirm deployed shape.	LEO, DragNet-1 deployed Nov. 2013, DragNet-2 planned to deploy in late 2017	[22, 23]
EGS Reflector	1999	Demonstrate deployment in micro-gravity.	• Image the deployment.	5-m-diameter pantograph hoop truss reflector with tensioned ribs and mesh aperture, integrated antenna reflector.	Video	Few orbits	[24]

Inflatable Antenna Experiment (IAE)	STS-77 May 1996	Demonstrate high payoff technology for large, inflatable space antenna structures, in a realistic operational environment.	• Verify that large inflatable space structures can be built at low cost. • Show that large inflatable space structures have high mechanical packaging efficiency. • Demonstrate that this new class of space structure has high deployment reliability. • Verify that large membrane reflectors can be manufactured with surface precision of a few millimeters' rms. • Measure the reflector surface precision on orbit.	14-m-diameter deployable offset parabolic with reflector membrane structure. The reflector structure consisted of 62 aluminized Mylar gores that were 7 microns thick. Deployment is enforced by an inflation gas injected into struts and the aperture torus.	Spring-loaded ejector panel that would push the stowed structure away from the canister. High-resolution cameras, digital imaging radiometer for measuring reflector surface pressure.	One orbit	[25, 26]
Interplanetary Kite-craft Accelerated by Radiation Of the Sun (IKAROS)	H-IIA 202 May 2010	Demonstrate the functionality of a rotational deployable solar power sail using centrifugal force as a deployment mechanism.	• Expand solar power sail and obtain expanded sail dynamics. • Evaluate performance of thin, flexible solar arrays attached to the solar sail. • Demonstrate the plausibility of navigation through photon pressure on sail. • Estimate a length and direction of acceleration vector of photon pressure.	Square solar power sail 14 m on each side with a cross-section of 20 m after deployment. Deployment is enforced by centrifugal forces.	The sail deployment mechanism consists of tip mass separation, relative rotation, rotation guide, and launch lock mechanisms. 80 images taken using a separation camera system consisting of a separation mechanism, image receiver, communications antenna, and analogue radio camera part.	May 2010–June 2010, (Venus Transfer Orbit)	[27]

(*Continued*)

TABLE 7.1 OVERVIEW OF SPACE FLIGHT TEST CASES (*CONTINUED*)

Mission	Launch	Purpose	Objectives	Test Article	Instrumentation	Duration	Reference
Interferometry Program Experiment (IPEX I)	STS-80 Nov. 1996	Characterize the vibrational environment of the platform to determine if it would achieve broadband nanometer stability requirements.	• Characterize on-board dynamic disturbances. • Predict boom response levels for IPEX-II. • Determine sun/shade thermal snap effects. • Correlate disturbance levels to ground test/analysis.	A/S bus structure with thrusters, gyros, tape recorders, and telescopes.	12 accelerometers (Sundstrand QA-2000), load cells, thermistors, and shakers.	2 orbits	[28]
Interferometry Program Experiment (IPEX II)	STS-85 Aug. 1997	Characterize quasi-random thermal-structural dynamic disturbances at a precision truss.	• Quantify the micro dynamic stability of a potential structural element of the Origins structure. • Measure boom thermal snap and dynamic attenuation response (IPEX II only).	9-bay articulated ADAM mast: graphite/epoxy struts with steel fittings (launched pre-deployed).	24 accelerometers (Sundstrand QA-2000), 8 load cells, 42 temperature sensors, and 2 shakers.	25 orbits	[28]
Joint Damping Experiment (JDX)	STS-69 Sept. 1995	Understand how the damping in a pin-clevis joint changes in micro-gravity by measuring the influence of gravity on the structural damping on a three-bay truss.	• Characterize the influence of gravity and joint gaps on structural damping and dynamic behavior of a small-scale truss model. • Evaluate the applicability of low-g aircraft test results for predicting on-orbit behavior. • Evaluate the ability of current nonlinear finite element codes to model the dynamic behavior of the truss.	Test structure is constructed with struts and pin-joints to form a truss structure of three cubic bays. Each bay measures 8 in.×8 in.×8 in.	Torsion and bending accelerometers and magnets located in stainless steel tip masses. CR9000 data logger, six Kistler K-Beam capacitive accelerometers, and CR10 Controller.	Total of 30 twang tests performed in orbit as well as post flight testing that followed.	[29]

Large Deployable Reflector Experiment (LDREX)	Ariane-5G Dec. 2000	Validate design of LDR by observing in-orbit deployment behavior of partial and scaled LDR models.	• Evaluate the validity of the modularized test method and the accuracy of analytical prediction of in-orbit characteristics.	Seven half scale hexagonal modules comprise the aperture. Surface is gold-plated molybdenum mesh covered by a structural cable surface with adjustment ties back to a secondary cable network (behind the mesh). The secondary network is anchored to several radial arms, each deployed as a four-bar linkage.	N/A	N/A	[14]
Large Deployable Reflector Experiment (LDREX-2)	Ariane-5ECA Oct. 2006	Confirm the adequacy of the LDR design modifications, and evaluate the accuracy of analytical predictions of in-orbit characteristics.	• Evaluate the validity of the modularized test method and the accuracy of analytical prediction of in-orbit characteristics.	Seven half-scale modules of LDR	Two video-motion cameras to observe deployment behavior. Deployment cable tension and deployed angle were measured.	N/A	[13]
Large Deployable Antenna Reflector (LDR) on Engineering Test Satellite (ETS-VIII)	H-IIA Dec. 2006	Evaluate the accuracy of analytical predictions of in-orbit characteristics.	• Establish an advanced 3-ton-class satellite bus technology. • Develop large deployable antenna reflectors. • Demonstrate mobile satellite communication system technology that will enable audio/data communications with handheld terminals. • Study basic technology of geostationary satellite positioning.	Two 19 m × 17 m gold molybdenum mesh reflectors S-band. 14 5-m hexagon modules.	Video-motion cameras to observe deployment behavior. Deployment cable tension and deployed angle were measured.	GEO, unknown duration	[12]

(Continued)

TABLE 7.1 OVERVIEW OF SPACE FLIGHT TEST CASES (*CONTINUED*)

Mission	Launch	Purpose	Objectives	Test Article	Instrumentation	Duration	Reference
LightSail-1	AFSPC-5 May 2015	Demonstrate deployment.	• Collect images of deployment.	Four 4-m stainless steel Triangular Rollable And Collapsible (TRAC) booms tension four Mylar film quadrants forming a square 32 m^2 planar area.	On-board camera.	356 km × 701 km orbit, launched 20 May, deployed 7 June, re-entry 14 June	[30, 31]
LightSail-2	STP-2 2018 (planned)	Demonstrate deployment and solar pressure thrust.	• Collect images of deployment. • Measure solar pressure thrust.	Four 4-m stainless steel TRAC booms tension four Mylar film quadrants forming a square 32 m^2 planar area.	Visible camera(s) from nearby spacecraft.	720 km altitude (planned)	[32]
Mars Advanced Radar for Subsurface and Ionospheric Sounding (MARSIS)	ESA's Mars Express Spacecraft, June 2003	Search for evidence of subsurface water and perform measurements on the Martian ionosphere.	• Successfully deploy above Martian atmosphere. • Gather data for deployment characteristics in order to develop and validate a better dynamic model for carpenter hinged deployment mechanisms.	Three foldable flattenable tubes (FFT)-TM constructed from Kevlar® and fiberglass composite tubes with periodic lenticular hinges. Two of the 20-m MARSIS booms form a 40-m dipole antenna, with the third boom acting as a 7-m monopole antenna.	N/A	Entered Mars' orbit on Dec. 2003	[33]
Middeck 0-gravity Dynamics Experiment (MODE)	STS-48, Sept. 1991	Study suspension and gravity influences on the structural dynamics of a modular truss system, by comparing the measured response in	• Perform several complete 1-*g* modal tests using suspensions of different stiffness and mass. • Conduct modal tests in 0-*g* during a space shuttle mission. • Quantify the suspension- and gravity-induced perturbations to the	A versatile set of modules allowing for different configurations. It included deployable truss modules, erectable truss hardware, a rotary joint, rigid appendages, and a flexible appendage.	Accelerations were sensed by piezo-resistive accelerometers, deformations by foil strain gauges, and input force by a load cell.	All scheduled tests were performed during two days of testing on STS.	[34]

		ground and orbital tests.	dynamic behavior using analytical models developed of the suspension and nonlinear effects.				
NanoSail-D (NSD)	Minotaur IV, Nov. 2010	Demonstrate sails as a means of nanosatellite de-orbit requirement compliance.	• Eject a nanosatellite from a microsatellite. • Deploy the sail from a highly compacted volume and validate de-orbit capability.	3U CubeSat with deployable 10 m^2 sail, four 2.2-m stainless steel TRAC booms.	Sail Activation Interface Board (SAIB)- UAH	7 months starting at 650 km on Jan. 20, 2011	[35, 36]
Rigidizable Inflatable Get-Away Special Experiment (RIGEX)	STS-123 March 2008	Demonstrate in space the stowage, deployment, and rigidization techniques of carbon fiber composite inflatable rigidizable booms.	• Demonstrate and validate deployment and rigidization techniques for three 50.8-cm-long (20-in.-long) carbon fiber booms (tubes) in a microgravity environment. • Test modal characteristics on orbit.	Three 20-in.-long inflatable rigidizable booms mounted to a sub-frame assembly that fits within the CAPE canister platform.	Pressure sensors, accelerometers, and temperature and voltage sensors. macrofiber composite (MFC) embedded actuators.	Single mission deployment	[37, 38]
Roll Out Solar Array (ROSA)	CRS-11 Jun 2017	Characterize the performance of a Roll-Out Solar Array in a space environment.	• Characterize deployment loads and kinematics. • Characterize deployed dynamics and the changes in full-sun and shadow. • Characterize blanket survivability and PV performance post deployment. • Characterize retraction kinematics.	5.36-m by 1.67-m wing including two high strain composite slit tube columns supporting a flexible tensioned blanket.	9 accelerometers (SD 2220-02 and 2270-02), 21 RTDs, 8 limit switches, 110 flat photogrammetry targets, and 4 spheres.	7 days, 108 orbits on ISS in LEO	[39, 40, 41, 42]
Solar Array Flight Experiment (SAFE) (also referred to as OAST-1)	STS-41D Aug. 1984	Characterize deployment, retraction, and deployed dynamics of a	• Demonstration of the existence of large area, lightweight solar array technology for SEPS. • Demonstrate array	31.5-m by 4.0-m accordion folded flexible blanket supported by a three-longeron glass coilable lattice truss.	Micro-g servo accelerometers in tip box, 23 retroreflector targets on blanket,	10 orbits LEO	[43]

(Continued)

TABLE 7.1 OVERVIEW OF SPACE FLIGHT TEST CASES (*CONTINUED*)

Mission	Launch	Purpose	Objectives	Test Article	Instrumentation	Duration	Reference
		flex-blanket array.	structural integrity for shuttle environments, full and partial retraction, and automatic restowage. • Demonstration of proper array performance in the thermal, vacuum, and zero-gravity environment of near earth space. • Correlation of electrical, thermal, and dynamic performance predictions. • Investigate the safe, reliable, and cost-effective experiment performance.		and retroreflector corner cube(s).		
Shuttle Radar Topography Mission (SRTM)	STS–99 Feb. 2000	Generate a near-global high-resolution database of the earth's topography.	• Survey a topographic database of 80 percent of the earth's land surface. • Successfully deploy 60-m mast and demonstrate functionality of mast vibration damping system.	60-m articulated graphite ADAM mast supporting a C-band synthetic aperture radar. Two radar antennas separated by a fixed distance to form an interferometer. The in-board antenna, relative to the Orbiter payload bay, is used as both a transmitter and receiver, while the outboard antenna is a receiver.	Astros Target Tracker (ATT) and three Optical Target Assemblies (OTAs). ATT was a star tracker that had been focused to 60 m. LEDs were mounted to the tip of the deployable mast, and the ATT was utilized to track its motion.	11 days LEO	[15, 44]

TABLE 7.2 FREQUENCIES AND MODE SHAPES OF NATURAL VIBRATION

Mission	Actuator	Instrumentation	Sampling Rate	Noise Floor	Analysis Method	Reference
ACTEX I/II*	Active tripod with integrated Navy Type II piezo sensors, external mechanical shaker	Endevco 7703A 1000 pC/g PZT accelerometers at top plate	1–1000 Hz operating range (No specific rate given)	n/a	Open- and closed-loop response of the tripod top plate response due to mechanical shaker.	[20]
IPEX I/II	2 proof-mass mechanical shakers, broadband random input tests, narrowband step-sine sweeps	24 micro-g accelerometers (Sundstrand QA-2000), 8 load cells, and 42 temperature sensors	1 KHz w/ 16-bit accuracy (IPEX I), 909 Hz	22 μg (IPEX I/II)	229 minutes of acceleration time histories (IPEX I). Analysis using pre-flight calibration information (IPEX II)	[28]
JDX	4 linear actuators	6 Kistler K-Beam capacitive accelerometers, Campbell Scientific CR9000 data logger	3000 Hz	n/a	Free-decay response	[29]

(Continued)

TABLE 7.2 FREQUENCIES AND MODE SHAPES OF NATURAL VIBRATION (*CONTINUED*)

Mission	Actuator	Instrumentation	Sampling Rate	Noise Floor	Analysis Method	Reference
MARSIS	Deployable dipole and monopole booms with integrated lenticular hinges	Mars Express rate gyros	8 Hz	n/a	ADAMS model compared to on-orbit deployment data.	[33]
MODE	Proof mass actuator shaker	Accelerometers, foil strain gages, and load cell	500 Hz	n/a	Half-power estimates of natural frequency and damping along with time histories converted to the frequency domain form.	[34]
RIGEX	5-cm-long macrofiber composite embedded piezo actuators mounted near fixed end	Triaxial accelerometers attached to the free end cap along with laser vibrometer, two K-type thermocouples, accelerometers	5000 Hz	n/a	Five minutes of forced vibration response testing.	[37]
ROSA	Motor and linear ball screw with self-centering flexures, narrow-band sine sweeps	9 micro-g accelerometers (Silicon Designs 2220-02 and 2270-02) in tip mandrel, blanket	Accelerometers at 200 Hz	8 μg but much higher near excitation motor	FFT analysis of the tip mandrel and blanket accelerometer response. Photogrammetry	[39, 40, 41, 42]

		mid-span, and root structure; 110 photogrammetry targets on blanket, tip mandrel, booms, and root structure; 4 external ISS cameras			analysis of blanket and tip mandrel motion.	
SAFE (OAST-1)	Dynamic excitation supplied by Orbiter Vernier attitude control system (VCS)	Data acquisition system (DAS) with tape recorder and video camera	n/a	n/a	Free-decay response using photogrammetry in day and laser tracker at night	[43]
SRTM	Two of the Orbiter's +X Primary Reaction Control System jets. Cold gas thruster mounted to tip of the mast.	ASTRO's Target tracker refocused to 60 m along with LEDs used for mast structural identification pulse tests.	n/a	n/a	ATT and LEDs used as a metrology system to track mast motion.	[44]

*ACTEX I/II information is based on planned measurements. Actual on-orbit results are not available due to the launch failure of DoD STEP M3 Spacecraft.

TABLE 7.3 CRITICAL DAMPING

Experiment	Actuator	Instrumentation	Sampling Rate	Noise Floor	Analysis Method	Reference
IPEX I/II	2 proof-mass mechanical shakers, broadband random input tests, narrowband step-sine sweeps	24 micro-g accelerometers (Sundstrand QA-2000), 8 load cells, and 42 temperature sensors	1 KHz w/16-bit accuracy (IPEX I), 909 Hz	22 μg (IPEX I/II)	229 minutes of acceleration time histories (IPEX I). Analysis using pre-flight calibration information (IPEX II).	[28]
MARSIS	Deployable dipole and monopole booms with integrated lenticular hinges	Mars Express rate gyros	8 Hz	n/a	ADAMS model compared to on-orbit deployment data.	[33]
MODE	Proof mass actuator shaker	Accelerometers, foil strain gages, and load cell	500 Hz	n/a	Half-power bandwidth estimates of natural frequency and damping along with time histories converted to the frequency domain form.	[34]
ROSA	Ball screw motor, narrow-band sine sweeps	9 micro-g accelerometers (Silicon Designs 2220-02 and 2270-02)	200 Hz	8 μg	Log decrement analysis of free-decay response of accelerometers in time domain.	[39, 40, 41, 42]
SAFE (OAST-1)	Dynamic excitation supplied by Orbiter Vernier attitude control system (VCS)	Data acquisition system (DAS) box and DAS tape recorder, TV camera		n/a	Free-decay response using photogrammetry in day and laser tracker at night.	[43]

TABLE 7.4 THERMALLY INDUCED DYNAMIC DISTURBANCES

Experiment	Actuator	Instrumentation	Sampling Rate	Noise Floor	Analysis Method	Reference
ACTEX I/II*	Nitinol wire wrapped around the flexure and top plate that contracts when resistive heating increases.	7 independent thermistors to monitor hardware temperature and gradients	6 thermal cycles across temperatures between −157°C and 148°C.	n/a	n/a	[20]
ROSA	Natural low earth orbital transition from solar night to solar day and solar day to night.	9 micro-g accelerometers (Silicon Designs 2220-02 and 2270-02) in tip mandrel, blanket mid-span, and root structure; 4 station cameras and 4 on-board cameras at the root tube	Accelerometers at 200 Hz. Response was measured for 10 solar transitions. 1 minute before and 5 minutes after daybreak	8 μg	No measurable change in accelerometer output was observed above the noise floor.	[39, 40, 41, 42]

TABLE 7.5 DEPLOYMENT FORCE AND TORQUE

Experiment	Actuator	Instrumentation	Sampling Rate	Analysis Method	Reference
MARSIS	Deployable dipole and monopole booms with integrated lenticular hinges	Mars Express rate gyros.	8 Hz	ADAMS model compared to on-orbit deployment acceleration data.	[33]
ROSA	Stored strain energy in carbon composite slit tube booms: 10 cm diameter and 535 cm long.	2 rotational dampers on either end of tip mandrel. 1 RTD in each damper.	1 Hz	During deployment, dampers generate a voltage proportional to the torque applied and the instantaneous temperature of the damper when compared to ground test results of the damper. This is not a direct measurement of deploy torque but rather the deploy torque minus all parasitic torques. It is more of a measure of deploy torque margin.	[39, 40, 41, 42]

*ACTEX I/II information is based on planned measurements. Actual on-orbit results are not available due to the launch failure of DOD STEP M3 Spacecraft.

7.5 SUMMARY

Upon review of the best space flight test cases in this chapter, several attributes emerge:

- Sharp, actionable flight objectives
- Early development of specific on-orbit measurements and pre/post-flight data analysis plans
- Baseline predictions of on-orbit behavior with analytical models that have been previously vetted against ground test data
- Clear understanding of the flight purpose and the risk posture by all stakeholders. Is it a test, an experiment, or a demonstration?

Each of the four case studies provides us additional best practices. For example, the LDR showed the benefits of multiple dedicated structurally focused flight tests prior to a mission demonstration flight. The first flight test, LDREX, had the simple objective to evaluate the validity of the modularized ground test method and the accuracy of analytical structural predictions. Two video cameras observed deployment while cable tension and deployment angle were measured. Excellent camera images and well-placed sensors on the first flight, plus ground test and analysis predictions, provided adequate forensic sources to determine the root cause of the deployment failure. Six years later, a second reflector, LDREX-II, was launched with the same objective, albeit an improved launch restraint and mesh deployment management system, and this resulted in a fully successful deployment. The lessons learned from LDREX and LDREX-II culminated in a full-scale LDR demonstration mission that included non-structures-related objectives: 1) demonstrate mobile satellite communications and 2) study geostationary satellite positioning.

The SRTM showed the importance of ground testing for very long 60-m structures and the approaches needed to measure the precision and structural dynamics. Flight objectives included 1) deploy a 60-m mast and 2) demonstrate the mast vibration damping system. The flight also included a clear operational objective to survey a topographic database of 80 percent of the earth's land surface. Therefore, deployment success was critically tied to mission success, a risk posture that certainly drove up ground testing costs. This program made extensive use of gravity off-loading techniques to test deployment on the ground prior to launch. Since the ADAM mast deploys in a linear direction without additional degrees of freedom, the ground deployment was effective at reducing the on-orbit deployment failure risk. This risk mitigation approach is likely not feasible for more complex deployment schemes such as that used on the LDR antenna, the ambitious SRTM mission was a great success due in part to precisely crafted mission objectives and extensive ground testing of the mast.

The SAFE demonstrated the feasibility of two important on-orbit test tools: 1) forced structural dynamic excitation and 2) photogrammetry for structural dynamics frequency and mode shape measurements. In this brief 15-hour mission, a flexible blanket structure was excited multiple times in both orbital day and night. In-plane and out-of-plane free-decay response was captured using an optical metrology system, showing the importance of measuring structural dynamics behavior in both extreme hot and cold, especially for a tensioned structure where unexpected blanket curvature and mast twist was observed during eclipse. Another surprise was that the first-mode critical damping ratio was measured to be much higher than the expected: 3.5 percent instead of 0.5 percent. These two observations highlight the importance of measuring structural damping on-orbit and the need to test dynamics in both solar day and night. Ultimately, the SAFE experiment significantly added to the community's flight measurements toolbox, including forced excitation, photogrammetry, and laser diode metrology during solar night.

ROSA demonstrated the importance of establishing a baseline prediction of flight performance through ground testing and modeling. This baseline data set proved valuable for quickly interpreting real-time flight dynamics data and adjusting the experiment plan accordingly. A mass and stiffness representative Engineering Development Unit (EDU) was constructed one year prior to the flight unit and was modal tested in a vacuum environment with the same actuator and accelerometer arrangement planned for the flight unit. This testing was critical to developing the excitation levels and durations for flight test cases. This test data was also used to train the analytical tools later used to quickly evaluate the wing dynamics data during flight operations. After vacuum dynamics testing the EDU was subjected to dozens of ambient deployment tests, was used to refine the flight wing gravity off-load approach, and was used to develop the flight wing assembly procedures. Yet, one of the regrets was not deploying the EDU in a thermally relevant environment as this would have identified a composite material batch issue in the STEM booms that manifested during ground functional tests of the flight wing. The issue was resolved but at a greater cost and schedule burden than if discovered on the EDU. Nevertheless, establishing the flight performance baseline prior to launch was essential to the success of ROSA.

The ROSA experiment also shows us the importance of securing partnerships and advocates in the flight experiment process and to expect long technology development timelines for these large structures. In this case, a full decade passed between concept formulation and commercial transition. The flight experiment itself was a four-year endeavor that required partnerships among AFRL, SMC, two NASA Centers, and four contractors. At the time of going to press ROSA is well on its way to being offered as a commercial solar array product by Space Systems Loral.

REFERENCES

[1] Tafazoli, M., "A Study of On-orbit Spacecraft Failures," *Acta Astronautica*, Vol. 64, 2009, pp. 195–205.

[2] Selding, P., "Launch of SiriusXM Satellite Delayed by Solar Array Concerns," *Space News Online*, 1 March 2012, http://www.spacenews.com/satellite_telecom/120301-launch-siriusxm-sat-delayed.html [retrieved 7 March 2012].

[3] Selding, P., "SkyTerra 1's Antenna Nearly Fully Deployed," *Space News Online*, 10 Dec. 2010, http://www.spacenews.com/satellite_telecom/101210-skyterra-antenna-deployed.html [retrieved 14 Feb. 2012].

[4] Kennedy, F., "Space and Risk Analysis Paralysis," *Aerospace America*, Vol. 40, No. 10, pp. 29–34.

[5] Wada, B., and Lou, M., "Pre-Flight Validation of Gossamer Structures," 43rd AIAA Structures, Structural Dynamics, and Materials Conference, Denver, CO, AIAA-2002-1373, 22–25 April 2002.

[6] Phone interview with Dr. Costa Cassapakis by Dr. Chris Jenkins and Dr. Jeremy Banik, 29 April 2016.

[7] "Schedule and Pricing," Spaceflight.com, http://www.spaceflight.com/schedule-pricing [retrieved 15 April 2017].

[8] Charania, A.C., "LauncherOne: Virgin Galactic's Dedicated Launch Vehicle for Small Satellites," 31st AIAA/USU Conference on Small Satellites, 8–13 Aug. 2016, SSC16-II-02.

[9] "The Annual Compendium of Commercial Space Transportation: 2016," produced for FAA AST by The Tauri Group under contract, pp. 17–19, https://www.faa.gov/about/office_org/headquarters_offices/.../2016_Compendium.pdf [retrieved 15 April 2017].

[10] Braun, B., Sims, S. M., Reese, K., Butzin, D., and Voss, D., "Navigating the Policy Compliance Roadmap for Small Satellites," Proceedings of the AIAA/USU Conference on Small Satellites, Launch, SSC16-II-6, 2016.

[11] "About Elana," https://www.nasa.gov/content/about-elana [retrieved 15 April 2017].

[12] Meguro, A., Shintate, K., Usui, M., and Tsujihata, A., "In-Orbit Deployment Characteristics of Large Deployable Antenna Reflector Onboard Engineering Test Satellite VIII," 58th International Astronautical Congress, Hyderabad, India, IAC-07-C2.2.02, 24–28 Sept. 2007.

[13] Meguro, A., Ishikawa, H., and Tsujihata, A., "Study on Ground Verification for Large Deployable Modular Structures," *AIAA Journal of Spacecraft and Rockets*, Vol. 43, No. 4, pp. 780–787.

[14] Meguro, A., Ishikawa, H., Tsujihata, A., Miyasaka, A., and Nakamura, K., "Analysis and Test Methods for Large Deployable Space Structures," 42nd AIAA Structures, Structural Dynamics, and Materials Conference, Seattle, WA, AIAA-2001-1479, 16–19 April, 2001.

[15] Duren, W., Wong, E., Breckenridge, B., Duncan, C., Tubbs, E., and Salomon, P., "Metrology, Attitude and Orbit Determination for Spaceborne Interferometric Synthetic Aperture Radar," Jet Propulsion Laboratory, California Institute of Technology, Pasadena, SPIE Conference of Acquisition, Tracking and Pointing XII, Orlando, FL, SPIE Vol 3365, April 1998.

[16] Schock, R., "Solar Array Flight Dynamic Experiment," *NASA Technical Paper 2598*, May 1986.

[17] Spence, B., and White, S., "Directionally Controlled Elastically Deployable Roll-Out Solar Array," United States Patent 8,683,755, 1 April 2014.

[18] "International Space Station Research and Technology, Roll Out Solar Array (ROSA)," https://www.youtube.com/watch?v=1Pm4MbIwRDw [retrieved 21 June 2017].

[19] Hoang, B., Spence, B., White, S., and Kiefer, S., "Commercialization of Deployable Space Systems' Roll-Out Solar Array (ROSA) Technology for Space Systems Loral (SSL) Solar Arrays," Proceedings of the 2016 IEEE Aerospace Conference, IEEE Conference Publications, pp. 1–12.

[20] Nye, T. W., Manning, R. A., and Qassim, K., "Performance of Active Vibration Control Technology: The ACTEX Flight Experiments," *Smart Matter. Struct*, 8 (1999), PII: S0964-1726(99)07557-6, pp. 767–780.

[21] Bushnell, G. S., and Becraft, M. D., "Flight Test of an International Space Station Active Rack Isolation Prototype System," Boeing Phantom Works, Kent, WA, 17 Aug. 1999.

[22] Robinson-Avila, K., "Keeping Outer Space Clean," *Albuquerque Journal Online*, 17 Aug. 2016, http://www.abqjournal.com/828114/keeping-outer-space-clean.html.

[23] Peterson, M., and Murphey, T., "Large Deformation Bending of Thin Composite Tape Spring Laminates," 54th AIAA Structures, Structural Dynamics, and Materials Conference, Boston, MA, AIAA 2013-1667, 8–11 April 2013.

[24] Klooster, V., Scialino, L., Cherniavski, A., Medzmariashvili, E., Korneev, V., Kravchenko, Y., Gulyayev, V., Magjanov, R., Santiago-Prowald, J., Lubrano, V., Silvestrucci, F., Fei, E., and Raboso, D., "Large Deployable Antenna for Various Space Applications," 5th International Conference on Antenna Theory and Techniques, Kiev, Ukraine, pp. 70–73, 24–27 May 2005.

[25] Freeland, R. E., Bilyeu, G. D., Veal, G. R., Steiner, M. D., and Carson, D.E., "Large Inflatable Deployable Antenna Flight Experiment Results," *International Astronautical Federation*, IAF-97-1.3.01, 1997.

[26] Freeland, R. E., "Significance of the Inflatable Antenna Experiment Technology," AIAA-98-2104, 1998, pp. 2789–2796.

[27] Sawada, S. et al., "Mission Report on the Solar Power Sail Deployment Demonstration of IKAROS," 39th AIAA Structures, Structural Dynamics, and Materials Conference, Denver, CO, AIAA-2011-1887, 4–7 April 2011.

[28] Levine, M., "The Interferometry Program Flight Experiments: IPEX I&II," Proceedings of the SPIE Conference on Astronomical Interferometry, Kona, HI. SPIE Vol. 3350, March 1998, pp. 707–718.

[29] Folkman, S. L., Bingham, J. G., Crookston, J. R., Dutson, J. D., Ferney, B. D., Ferney, G. D., and Rowsell, E. A., "The Joint Damping Experiment (JDX)," *NASA Contractor Report No. 4781.* June 1997.

[30] Biddy, C., and Svitek, T., "LightSail-1 Solar Sail Design and Qualification," Proceedings of the 41st Aerospace Mechanism Symposium, Pasadena, CA, 16–18 May 2012.

[31] Chang, K., "LightSail Spacecraft Sends Back a Selfie Showing Its Sail Stretched Out," *New York Times* nline, 9 June 2015, http://www.nytimes.com/2015/06/10/science/

lightsail-spacecraft-sends-back-a-selfie-showing-its-sail-stretched-out.html?_r=2 [retrieved10 June 2015.

[32] Planetary Society Website, http://sail.planetary.org [retrieved 27 Jan. 2017].

[33] Adams, D. S., and Mobrem, M., "Deployment Analysis of Lenticular Jointed Antennas Onboard the Mars Express Spacecraft," *Journal of Spacecraft and Rockets*, Vol. 46, No. 2, March–April 2009.

[34] Barlow, M.S., and Crawley, E.F., "The Dynamics of Deployable Truss Structures in Zero-gravity: The MODE STA Results," *MIT Space Engineering Research Center, SERC* # 1–92, Jan. 1992.

[35] Whorton, M., Heaton, A., Pinson, R., Laue, G., and Adams, C., "NanoSail-D: The First Flight Demonstration of Solar Sails For Nanosatellites," 22nd Annual AIAA/USU Conference on Small Satellites, SSC08-X-1. 11-15 August, 2008.

[36] Alhorn, D. C., Casas, J. P., Agasid, E. F., Adams, C. L., Laue, G., Kitts, C., and O'Brien, S., "NanoSail-D: The Small Satellite That Could!" 25th Annual AIAA/USU Conference on Small Satellites, SSC11-VI-1. 8–12 August, 2011.

[37] Cobb, R. G., Black, J. T., and Swenson, E. D., "Design and Flight Qualification of the Rigidizable Inflatable Get-Away-Special Experiment," *Journal of Spacecraft and Rockets*, Vol. 47, No. 4, July–Aug. 2010, pp. 659–669.

[38] Black, J. T., Cobb, R. G., Swenson, E. D., and Cooper, B. J., "Rigidizable Inflatable Get-Away Special Experiment Space Flight Data Analysis," *Journal of Spacecraft and Rockets*, Vol. 48, No. 3, May–June 2011, pp. 477–487.

[39] Chamberlain, M., Kiefer, S., Jones, T., and Banik, J., "On-orbit Structural Dynamics Performance of the Roll-Out Solar Array," 5th AIAA Spacecraft Structures Conference, 8–12 January, 2018. To Appear.

[40] Kiefer, S., Chamberlain, M., and Banik, J., "Structural Analysis Methods for the Roll-Out Solar Array," 5th AIAA Spacecraft Structures Conference, 8–12 January, 2018. To Appear

[41] Jones, T., Liddle, D., and Banik, J., "On-orbit Photogrammetry Analysis of the Roll-Out Solar Array," 5th AIAA Spacecraft Structures Conference, 8–12 January, 2018. To Appear.

[42] Banik, J., and LaPointe, M., "On-Orbit Validation of the Roll-Out Solar Array," *IEEE Aerospace 2018 in Big Sky*, MT, 4–11 March 2018. To Appear.

[43] Elms, R. V., Young, L. E., and Hill, H. C., "SEP Solar Array Shuttle Flight Experiment," Lockheed Missiles and Space Co., Sunnyvale, CA, 16th Intersociety Energy Conversion Engineering Conference, Atlanta, GA, 9–14 Aug. 1981.

[44] Umland, J., and Eisen, H., "STRM On-Orbit Structural Dynamics," 42nd AIAA Structures, Structural Dynamics, and Materials Conference, Seattle, WA, AIAA-2001-1588, 16–19 April, 2001.

APPENDIX 1

A central part of this volume is capturing the knowledge and experience of the technical experts who have been involved with the testing of this unique class of structures. Unfortunately, many of these men and women are at a situation in life that makes contributing in writing in a conventional manner difficult. Through the use of oral and written interviews, we have been fortunate to remedy this situation in the case of three experts: Costas Cassapakis, Robert Freeland, and Martin Mikulas. The interviews follow.

Date:	**April 29, 2016**
Interviewer 1:	**Chris Jenkins**
Interviewer 2:	**Jeremy Banik**
Interviewee:	**Costa Cassapakis**

BIOGRAPHY FOR DR. COSTAS G. CASSAPAKIS

Dr. Costas Cassapakis is a physicist with extensive work on precision inflatable space structures. He obtained a Ph.D. in High Energy Physics at Los Alamos in 1975, after which he joined Science Applications International Corporation to work on nondestructive testing of materials using nuclear radiation. In 1987, he became associated with L'Garde, Inc., to work on SDI-related programs, which involved the use of space inflatable systems for nuclear radiation testing. He is a past President of L'Garde, Inc., in Tustin, California.

INTERVIEW

Chris: Okay and we are recording this. And Costa, again, Jeremy and I both appreciate you taking a little bit of time here with us today.

Costa: Oh it's my pleasure really. It's really is. Yeah, I have a few things in my brain that may go to the younger generation. Hopefully we are going to avoid some errors that I made.

Chris: That's exactly right. I mean the whole purpose of this book is to try to capture a lot of this unique technology that we've developed over the years before it gets away from us.

Costa: Yeah, well I hope so because I got a whole bunch of examples that say that people come up with all these new ideas, not knowing that they are 50 and 60 years old. It's a good thing just from that particular viewpoint.

Chris: So Jeremy sent you I think four questions. We will just kind of start there. We will try to keep this to an hour.

Costa: I have all afternoon because Jeremy in the beginning told me that it was going to be about three hours. So I am not known for being very generous. But I will be more than happy to limit it to an hour.

Chris: Okay. Thank you. Tell us a little bit. I know you've had extensive background and go back very far in testing large ultra-lightweight structures and spacecraft. Tell us just a little bit about your background and your history in that area.

Costa: Well, I started out at SAIC, during the Star Wars era when we were trying to test the neutral particle beam weapon, which, it turned out, it couldn't happen. After that they turned it into a detector kind of a sensor in space whereby they would shoot neutral particles towards an incoming threat missile. And tried to get some ideas as to what that sensor detected from backscattered charged particles, neutral particles, gamma rays and such. But it turned out that we had a requirement for this to happen from a large distance and you can't get very much backscattering at 3 kilometers minimum distance.

So I got involved in the target vehicle of the experiment to come up with very large balloons. Some of them being decoys and some of them being loaded with heavy elements to see whether we could differentiate between decoy balloons and actual targets that may have had been impregnated with, say tungsten or tantalum to simulate uranium 238 and/or 235. So that's how I got started. And then a couple of years later I ended up working at L'Garde on an MDA program. It was called BMDO at that time... And I was the program manager for the so-called Firefly program which was fairly large. Not so large as high gain antennas and solar concentrators. But it was an inflatable that had certain material properties, and it also had numerous sensors on it. And something like about 32 channels communication to send back the data as we were collecting it after the launch Sandia National Laboratory was involved in it also. They provided the sub-orbital buses and each bus carried one of those targets as their only payload.

So we managed to launch two of those targets. We had built about three, I think, so after that, I got involved in the Inflatable Antenna

Experiment, which I worked with Tom Campbell of NASA Langley. And the management of that, as you all know, went to JPL under Robert Freeland. And the project lasted for about six years and we finally launched the Inflatable Antenna Experiment as the primary payload of the shuttle Endeavor in 1996. Another application I got involved in at the time was the Inflatable-Rigidizable Solar Array for satellites, A DARPA project, attempting to assess the role of aluminum foils rigidizable structures in space.

And that was really the beginning of this fascinating technology. There was extensive testing on all of those membrane articles. Both on the mechanical properties of the material (and that was the main thing, along with making sure we were operating in the linear portion of their modulus curve). But also the other subsystems involved in the process (like inflation systems, valves and such). So that's in a nutshell that I continued up till 1997 I think it was when I became the CEO of L'Garde. And we worked on a large number of large inflatable structures, mainly solar concentrators, and large gain antennas, and their lab models. And then we went into solar sails, and we built I think, probably the only existing large model that has been tested to date, it was 20 meters by 20 meters, which we tested for deployment at NASA"s, Plum Brook Environmental chamber. It was the only one this model could fit in.

And we did a few other things involved large membranes. And then after I left L'Garde I also contributed to a couple of programs. One was a high altitude airship. And the other was a large membrane mirror that would have deployed at a high earth orbit. It involved somewhat complex optics. So that's in a nutshell my experience with the technology.

Chris: Yeah. Good, thank you Costa. I wanted to go back to the, kind of your early experience with the decoys and that sort of thing. Did the company you work for initially or L'Garde afterwards have, at that time, any experience in testing these inflatables?

Costa: Oh yes. I joined L'Garde in 1987. The inflatable decoys were as an idea existed since 1965 when the Russians came up with the Galosh interceptor. And we had to devise ways to defeat it. So, the idea was that for every real missile we would send up a bunch of inflatable decoys. And the radars, both ground radars and airborne radars would have a hard time telling the difference between the real thing and the decoys. So that they wouldn't necessarily aim for our weapons.

Chris: But for example did you pressure test those decoys and try to measure skin strain to measure . . .

Costa: Oh absolutely.

Chris: How were some of those experiments or tests done?

Costa: Well, we would every so often, we would take the roll and cut 1 inch wide by 6 inch long samples from the roll that we got. And we would put them on a tensiometer and we would measure the stress and strain results both in the fill and the warp direction.

Chris: Yeah, sure, the transverse direction. Sure.

Costa: Transverse was the fill I think, yeah, and the warp is the long direction.

Chris: What was the material Costa?

Costa: Well, we had all kinds of materials. The cheapest one at the time was Mylar of course. We would test them mainly for modulus as I said in both the warp and the fill directions. And we would also check them optically to see what their optical characteristics were because we were trying to get the alpha over epsilon of those materials (i.e their optothermal properties). You needed to know what their thermal characteristics were, in order to do some thermal calculations find out for how long the entrained heat would be contained so that the decoys would have a similar thermal signature as the massive object we were trying to fly along with them.

Chris: Okay. I see.

Costa: Of course along with that you had to know what the density of the material was just so you could figure out what the mass of the material that you were going to use for the shape that you were flying was. So we had to measure that as well.

Chris: Did the shape have to remain fairly exact?

Costa: Yes, for those particular things that we did, the decoys, the shape had to be not as exact as you would want it, let's say for a solar concentrator or an antenna. I would say plus or minus a centimeter.

Chris: So did you look at viscoelastic aspect properties at all? Did you do any creep tests?

Costa: Whenever we had the material loaded, with elastic materials, yes, we still had to measure the modulus of that material. Furthermore the mission lifetime of the decoys was short most of the time and creep was not a serious concern. I was never involved in viscoelastic property testing. Other than trying to figure out what the modulus was. The most important thing is the modulus, if you are going to try to maintain shape and also not to burst the inflatable.

Chris: Okay. Jeremy, do you have any follow-up on any of that?

Jeremy: The one question I had Costa was regarding failure testing of these decoys. How often did you do it and how did you do it?

Costa: We would over-pressurize one object to burst. And we would get the pressure and the volume of gas that we put into it. That way we had a fairly decent idea of how the material was going to sustain all the loads.

Now, having said that, it brings me to another point I want to make. And that is that the modulus of the material and the uniformity of the modulus is an extremely important characteristic particularly as you go to large structures. And the reason is that you don't expect the non-uniformity of the material to show up if you have reasonably small objects, say a couple of meters or 3 meters. But when you get to large inflatables, then you really don't know the exact modulus at all points of the material. In a previous lifetime I tried to measure the thickness of the material with radiometric techniques like beta or gamma ray backscattering or x-ray backscattering. We found out that those were not linear and inaccurate.

In my opinion, this remains a concern. Although I think that you could possibly do it anymore with just lasers in the visible, which has a much lower frequency than the stuff that I was using. So this is the thing. You still run a probability some risk that your initial measurements may not be borne out by the bulk of the material. So it's very important particularly when you are dealing with large structures, to measure the modulus every maybe 5 or 6 meters worth of gore that you are trying to cut. Have I answered your question Jeremy?

Jeremy: Yeah, thank you.

Chris: So Costa the first company you went to work for, what was the name of that again?

Costa: It was SAIC, here in La Jolla. Science Applications International . . . Well in the beginning it was called Science Applications Incorporated. Then it became Science Applications International Corporation. Well, I think it went public a few years ago. SAIC separated itself into two companies called SAIC and Leidos Corporation. That's the company that I went to work for. I was a senior scientist there eventually, and associate director for radiometric gauging. So we would develop instruments that could tell us basically the content of certain elements, in processes like when we are making RDX or TNT type of explosives. And we wanted to know how much of each we had in a steel pipe. So we would punch the steel pipe real time with radioactive beams, mainly gamma rays, and sometimes neutrons. And we would then, knowing the cross section of interactions between these rays, and the relevant materials, we could from detecting them on the other side of the pipe, we could figure out what the densities involved, where and what the elements were. So that's what I was doing there mainly. And that's why I took part in the neutral particle beam experiments in Star Wars which brought me to L'Garde.

Chris: At SAIC, were you working with inflatables or membrane structures?

Costa: No. I was not.

Chris: So they weren't working in that area. It was when you went to L'Garde that you first started working on that.

Costa: That's correct.

Chris: How long had . . . when you went to L'Garde, how long had they been working with inflatable membrane gossamer technology?

Costa: At Ford-Aeronutronic, the engineers studied inflatable structures right around 1968. I think it was under a BMDO (Ballistic Missile Defense Organization) contract. It was BMDO and Aerospace Corporation. And then in 1971, they broke off because the decoys were not really a good business that Ford Aeronutronic wanted to pursue. They formed L'Garde in 1971. So I know for sure that they started working the decoys and the inflatables with the so-called Inflatable Exoatmospheric Object, IEO, which was made with a laminate of different materials including elastomers.

Chris: Okay, all right. Very good. Thank you. Second question is could you describe the situation when a test proved critical to the success of a program. Maybe by way of test you had a surprising result or you discovered an unknown, unknown, something you might not have found otherwise had you not actually done the test.

Costa: Yes. We had some articles fail at the seams since you have to seam those things, in order to achieve a certain accuracy of shape. We had objects, that when we first inflated them at atmosphere, they may fail at the seam and then you have to go and see how the seam failed. Was it because of insufficient level of adhesive applied or what the failure mechanism was? And fortunately we would get those questions answered, but not in vacuum chambers or else you would have a mess to clean up, and it would be expensive. But first you start testing these objects at laboratory environment, in order to find out that in fact they were built soundly.

I am going to bring you another example of that. I am not going to mention any names but when I was doing the Firefly target, the customer thought that we were being too expensive. And they decided to do the testing of the Firefly target in their own vacuum chamber. Now, as you know we use one shot mechanisms in order to time the particular events that have to take place like ejection for instance from the spacecraft followed by some other events, followed by inflation and target deployment. And I remember that very well.

The One-Shot event initiators were Pyro-electric devices. These have certain requirements, namely, how much current, how much voltage, how much resistance, must they use to function. And you know the whole process of firing a Pyro takes about a few milliseconds for the event to happen. Well, this particular engineer did not take into account the Pyro characteristics, and the inflation Pyro misfired before the containment pyro. And the whole target assembly got blown up in the vacuum chamber. And I am not exaggerating, but according to this engineer it took them about six months to clean up the vacuum chamber.

Chris: Great story.

Costa: You have to be extremely careful when you are working with things like that. To tell you the truth, when I went to L'Garde, I felt that hell, there couldn't be anything harder than experimental nuclear and particle physics. I went there thinking that what the hell is there in a balloon anyway. Man, was I proven wrong! It takes fantastic accuracy in all of your testing to make sure that things happen as they are supposed to, in the time they are supposed to, in the timeline that you have for all of those events. Because let's not kid ourselves: it's not just the inflatable. Once inflatable becomes a payload, it can endanger the whole spacecraft, if you don't have something light.

Another thing, that one must bear in mind for inflatable or deployable membranes Suppose you have a target or any kind of inflatable that goes up, in its container, and gets into a spacecraft bay. You've got to have an excellent bypass valve for the entrained air. Because when you are packaging it, you are also packaging air in it. Well, you have to allow a path for that air to escape during the launch before you can do it. Because before launch you are sitting at the launch pad and you have an atmosphere so the entrained air is not capable of escaping. But as you launch, and you go up to let's say lower orbit, you better make sure that that entrained air has a path for escape. That is normally done with a valve, which stays open, during launch, and then it shuts down when you reach altitude.

So, you have to allow for the entrained air to escape. If it doesn't, then the differential pressure inside the container of the inflatable increases to a dangerous point. And then when you try to inflate it, you over-inflate it, and you can burst it, even if it doesn't cause any damage to the rest of the spacecraft prior to ejection and deployment. You've got to be very careful for that. And that normally happens in very large inflatable structures. Because as you are folding those objects you are by necessity going to entrain a large amount of air.

And that is going to expand if you don't get rid of it. For instance in the inflatable antenna experiment which you are aware of, we had to wait for like a couple of days to make sure that all the gas that was in the folds of the antenna was able to escape. And, as it turned out, it wasn't enough.

The other thing that you can tell and I think it's very important is checking for pinholes in the material. Because thin materials will come from manufacturing with pinholes. The thinner the membrane the more pinholes you can expect and you need to know the average number of pinholes in order to establish the leak rate that you are going to have after deployment. Note that this is beyond the expected micrometeoroid damage. And so you have to account for the leak rate.

Chris: So Costa, how did you measure leak rate on a large inflatable, like [Inflatable Antenna Experiment] IAE? How did you find pinholes?

Costa: We, at the time did use balloons built from the same material and checked the leak rate in vacuum.by observation and analysis. But I had seen ILC [ILC Dover] taking rolls of the material like Mylar or Kapton. And they did it by inspection. It involved rotating the roll behind a fluorescent lamp. And then turning the crank manually. And as the roll unrolled or rolled they would check for pinholes. I don't think that's a good technique as human error can affect the results.

Also, don't forget, that, if you were to do this in the lab, the leak rate in vacuum is different. The leak rate tends to diminish when you are in vacuum. It's higher when you do it at atmosphere simply because you are inflating the balloon with much more air than it would take to inflate in vacuum. Namely, you have to go, if you need some kind of number of psi. Let's say 0.2 psi in vacuum. If you do the experiment in atmosphere, you've got to account for the 14.8 psi of the atmosphere. So you inflate the balloon to 15 psi. You get what I mean? It's not only the Delta P that you need to be aware of in the leak equations, but the number of air molecules producing the required Delta P has to be taken into account. And it turns out that if you use a balloon, in full atmosphere before you start looking for the effects, you are going to have the air escaping faster. Simply because the density now in the balloon is larger. So when you go into vacuum, that leak rate reduces fairly considerably. And that's how you can also determine analytically what the quantity of the makeup gas is that you have to send up with the inflatable.

Chris: Okay, very, very good. So third question we wanted to ask you, can you recall any highly unconventional or even bizarre test practices your team was forced to use, and how well they worked. I mean, did you have to borrow, bring your eggbeater from home?

Costa: Unconventional, I don't know what you mean by unconventional. Well, we used to do a lot of videogrammetry. In the beginning, before it was invented, we were using photogrammetry and it was very expensive. And that was to find out the exact curvature of the shape that we were trying to achieve. But I wouldn't call that unconventional. We were the first ones to use non-linear analysis in order to figure out the exact shape, let's say of a parabolic reflector. And we developed a code for that. But as you know now Abacus can do the same thing and it's commercially available. It has been for a number of years.

The other thing that I can think of is if you want to test your calculations and your results and your knowledge of the material properties, say you want to build a circular on axis antenna, you have to build a drum that you can attach the antenna on. And then you pull a vacuum on it equivalent to the pressurization that you are going to apply when it's in vacuum okay. So you pull a vacuum on it, and you get the concave surface that you are after achieving. And then you do videogrammetry to see that in fact at the exact pressure, you are going to get the shape you want. I don't think many people had realized that. But that's the cheapest way of finding out what your intended shape is before you build a canopy for it and fly the object.

That reminds me, during the inflatable antenna experiment, we had a drum that was 15 meters in diameter. It was an off-axis antenna. So the drum was not circular. It was slightly elliptical. Its height was about a meter. We had a 1 inch diameter hole on its side from which we would pull the vacuum on it, in order to measure the concave surface of the membrane. Well, after we did all the measurements and all that, we took the vacuum pump out of the 1 inch hole. And that antenna, it took about 3 days, 72 hours to where we started seeing wrinkles on it. So we had done our calculations right. Just think about that. That's a big volume. 15 meters round roughly by 1 meter high is the drum. And it took almost forever at atmospheric conditions for the leak rate, to get to the point where they destroyed the shape. That's one of the reasons we thought at the time that inflatable membrane antennas and concentrators made sense for space.

Another time we were pressured (no pun intended) to test all of our Pyros, the pyrotechnic devices that we used for sequencing deployment events. What they do is they build like a 1000 of them and then they take a fewfrom the lot. And they exploded them, just to see how well they've been built. And sometimes, they tell you that the batch they are sending you is good to within a 0,999 reliability. Thus, 1 out of a 1000 Pyros that they make is not going to work. But when you have such a critical timeline for the events to happen in your inflatable, you ought to make sure that there is no possibility for malfunction.

So what the bus people made us do is take them to a radiometric facility and subject them to neutron irradiation, so that we would get cold neutron pictures. It's sort of like x-raying the pyros. But x-rays cannot penetrate the Pyro. So they used neutrons and they would take pictures with the neutrons and then they see that in fact all the explosives is there. And that there is a connection, the wire between the explosives and the trigger is intact. I believe that that's a good idea if you can scare up the funds for it.

Chris: Yeah. But I am sure that in your experience, in this area, you had to cobble together probably a number of different tests techniques and test rigs and that sort of thing that just weren't available. And you probably had to figure them out on your own how to do those tests.

Costa: Well, for instance, when you try to deploy a flat membrane, you don't want the gravity effect to mess up your results, so you will have to devise ways to take it out. And yes, we would build a lot of those rigs. We would hang the membrane objects from wires that went into pulleys and the pulleys were running along a horizontally stretched wire, in order to affect zero gravity effect on the material. So I don't know if that's what's you are talking about.

Chris: No. That leads into really the last question, and I will just ask it, and see if Jeremy has any questions left, and see Costa if you want to make some final comments. What was the biggest testing need maybe back then or maybe it's still the biggest testing need now. What do you see future direction of testing of these large ultra-lightweight structures?

Costa: I think the biggest testing need, both literally and metaphorically, is a large thermal vacuum chamber. I would call that probably the biggest testing need that I've had because they are way too damn expensive to operate. And I've done a couple of those. In fact, one is probably next door to Jeremy. They have a very decent thermal vacuum chamber. Well, at least they used to. I don't know if they still do. And of course the apple of my eye was always the NASA thermal vacuum chamber at Plum Brook. I would say that that is probably for large structures, that is where you want to get at. For smaller structures, it just depends on what the dimensions of your object is going to be. Let's not forget that space environments, (thermal effects, micrometeoroids, radiation, atomic oxygen in low earth orbit, etc.) are very serious hazards for anything we put out there. The need to be determined as a function of orbit both analytically and, where possible, by testing.

But I think that is what I would consider the biggest testing need. The second to that one is the inspection of the material, and making sure that the material is very uniform if you expect to achieve a certain surface accuracy. And when you've cut the segments to form a parabolic or spherical surface, it's good to test the accuracy of cutting the membrane. We used to do this on a light table equipped with an optical comparator.

I think that the testing of the material properties is probably the next biggest thing that people need to pay particular attention to.

Chris: Jeremy, did you have anything else you wanted to ask Costa?

Jeremy: Sure. I wanted to circle back one more time on the Inflatable Antenna Experiment, simply because space flight tests are so rare in our business. There are some great lessons learned on IAE. What I wanted to ask you was, looking back on it now, are there any additional tests or additional things you would have done if you were to do it again? Maybe ground tests. Maybe it's one of the space flight test you would do differently.

Costa: Yes. What the Inflatable Antenna Experiment did, there were a couple of lessons to be learned there. The first lesson was because the structure was so large, we had to mount the uninflated structure in the spacecraft on a plate that would "kick" it forward to almost its full length before the inflation process would be started. I am talking about both the structure that supported it and the inflatable membrane itself. So we had to kick it out. Well, what happened, because of the entrained air, not having totally escaped, the inflatable unfurled almost half the way out. So when the plate was given the command to push, the folded inflatable, when it came out, it didn't see anything. So that didn't work. And so the inflation process started before the inflatable antenna had reached its end.

Do you understand what I am saying? Because I am saying that, I am waving here my own arms at it. And I don't have a blackboard to show you, just exactly what I mean. But if you start the inflation before you give the chance for the entrained air to get out, then you are going to get all kinds of unwanted results. And you will see from the video of the IAE that it really struggled to get to its final configuration. It created all kinds of unwanted loads. And then the other thing that I would have done differently is to check the connections of the support structure to the rest of the spacecraft. Make sure that those structures which in this case were also inflatable, were bundled right. In the case of the IAE if you remember from the video, it started rotating. And the reason for that was that there was a leak in one of the support arms that we figured out from the rate of rotation of the structure that it was right at where one of the inflatable arms was adhered to the solid structure that contained the inflatable. So those were things that I would have done differently. I mean I would have done much more testing, but neither funds nor schedule allowed it.

Jeremy: That's a great example. Did you guys do any entrained air testing on that and how?

Costa: Well, I don't think we did entrained testing in the folded configuration in the antenna spacecraft. I don't think so. What I would do differently is have some kind of a sensor just a very simple sensor to check what the pressure is in a couple of, three of the folds. And not start anything unless that pressure was very close to zero. And have that in the software that controls the whole experiment. And you can do that. You can put film transducers on the folded membrane or inflatable structure. And you can feed them out to the central computer, the central processor of the whole experiment. And they can measure in certain critical locations, what entrained air quantity there is. And so unless it reaches a critical point that you consider to be safe that's when you start the rest of your timeline. Don't forget it takes a very low differential pressure to get things inflated in vacuum.

Jeremy: Yeah, that's a great example, great explanation. Thank you.

Costa: In summary, gentlemen, as you may have realized, this is an inexhaustible subject and I'm sure we have not addressed all the issues associated with this technology, but hopefully it will help your readers recognize some of its complexity and take it into account very seriously.

APPENDIX 2

Date:	**September 6, 2016**
FILE NAME:	**Robert Freeland Interview**
Interviewer 1:	**Case Bradford**
Interviewer 2:	**Greg Davis**
Interviewee:	**Bob Freeland**

ROBERT FREELAND BIOGRAPHY

Mr. Freeland holds a BS in Mechanical Engineering from the University of Arizona, 1957, and an MS in Engineering Mechanics from University of Southern California, 1964. He has worked at JPL since 1958, where he has advanced technology of large mechanical space antenna structures (he has more than 70 referenceable publications on the subject). He has developed designs and scenarios for the orbital mechanical characterization of deployable antennas space shuttle based, space station based, and on free-flying spacecraft. Mr. Freeland managed the development of a 55-meter diameter deployable wrap-rib antenna structure. He was the Project Manager for the NASA Inflatable Antenna Experiment. Mr. Freeland is the recipient of the NASA Exceptional Engineering Achievement Gold Medal and the NASA Exceptional Achievement Gold Medal.

Interviewer 1: Alright. This is Case Bradford with Greg Davis and Bob Freeland on the 6th of September.

Interviewer 2: For posterity.

Interviewer 2: Okay Bob, so, let's start off with the first question and given that this handbook is about testing of lightweight space structures, could you give us the benefit of your considerable experience on the role of testing in the development and build of lightweight structures?

Bob: Specifically, what concepts I [worked on]?

Interviewer 2: Well, more along the lines of so, you have a conceptual design in mind, how does the test program fit in with that conceptual design. What levels do you test at, you test at component level, do you test at subsystem level, you test at...

Bob: Well, what you can do in the arena of structural testing, verification, in particular in deployables, the key feature is deployment reliability. Doesn't matter how good it is, doesn't matter what it costs, doesn't matter what the folding ratio is, ratio of functional dimensions compared to stored, doesn't matter. Everything is deployment reliability and that is the reason [for testing] from the military's point of view. They were very interested and bought a lot of the Lockheed Wrap-Rib antenna, going all the way back to the ATS-6.

Interviewer 2: Right, okay.

Bob: And the reason that is so good, wrap-rib, it has only got one real mechanism in it and that's the hinge of the rib, as they pulled up elastically around the hub and the mesh was tied very tightly connected to the rib so when the antenna is open, the effective value of the parabolic precision is called the flats approximation of the mesh between each parabolic rib. Now, if you really want to get down to details, there is a slight pillowing effect because the tension in the mesh between the ribs is slightly different in each direction, but that really is a third-order effect, so you can forget that and typically when the DOD order an antenna for a certain wavelength, they work backwards to find out how many ribs do they need to come up with a flats approximation for a certain percent efficiency on axis and so what that boils down to is how big a one can you test on the ground, so the scaling factor is very small and it turns out that's very hard to do because the real big antennas are quite floppy because if they weren't if they were stiff enough to do meaningful things on the ground, they would be so heavy, you couldn't lift them and the package volume would be so large, it would be a major, major problem.

Interviewer 2: So, Bob, let me harp in for a minute. These are all great touch points, when you say deployment reliability, are you talking about...

Bob: Kinematic!

Interviewer 2: Sorry talking about just the first motion test or are you talking about a full deploy. Are you doing this ambient or are you doing this in vacuo.

Bob: I am talking about the reliability to its operational configuration on orbit from the stowed package configuration.

Interviewer 2: Okay, so you are talking about the full deployment.

Bob: That's it. That is the most important and that's the word I get from the people that have used large amounts of these.

Interviewer 2: Yes, so are 1G ambient conditions sufficient to demonstrate deployment reliability or would you prefer to do this at temperature in a vacuum? Are those significant factors?

Bob: Of course the latter would be better, but these wrap-ribs are, they are really built well, they are really very rugged and the hinge has only got bending loads on it, it doesn't have a lot of torque loads on it and so they are quite rugged. The enemy of ground demonstration is the gravity deflection. Now, what do they do about that to try to get around it? They build mathematical models to predict the deflections you are going to get under 1G in the small size range. What's a small size range: 5 to 10 meters. They figure that's about as big as you can go. I built, I didn't but Lockheed built on a contract, I managed for NASA, technology for a 55 meter diameter reflector for science applications, but in the woodwork the military who had used smaller were very interested in this too.

Interviewer 2: So, let me go to the 0G thing for a minute.

Bob: What?

Interviewer 2: The 1G for a moment, so you do the mathematical analysis to predict your deflections under IG loading, but of course when we do ground demos, we often build an offloading fixture, right?

Bob: Right.

Interviewer 2: And what kinds of problems do those introduce into the test?

Bob: Okay, the biggest one they built previous to my involvement was the ATS-6 and tested that on the ground, but the objective from NASA and other financial supporters of that activity was one at least 50 meter diameter, so we selected 55 meter diameter based on the largest space Lockheed could test it and have a roof high enough to support the 0G stimulation, so what we did, we built a sector, I talk about in several papers I wrote, I can get for you. This is well-known information.

Interviewer 2: We can list those as references that would be great.

Bob: So, what we did, we built 4 ribs in pieces and spliced them together and then we machine the parabolic contour. Now, the first question, how do you get the parabolic contour, very simple, we put it in a floatation tank and that is the same rough radius of curvature and we float it and then from the unloaded quote "floatation configuration", we use a laser to mark where we machine it, okay, so after the ribs are fabricated, put together, we machine it. We put the hinge in. Now, the objective is to have something with this diameter that is as high mechanical packaging efficiency which means you don't want big thick or deep ribs to minimize deflection because at 0G they don't really care, so we selected a design actually based on a combination of what we want in orbit and to enable ground testing. If you make a rib like tissue paper, which they could do in space if we could get it up there and open it without damaging them, that would work. It was a combination of the smallest thing, we could test on the ground that would end up in reasonable package efficiency and weight. So, then, above that.

Interviewer 2: So, Bob, let me hop in here for a second, so actually and we hear this a lot, the ground environment in some regards is the design driver for constructing these large antennas because they have to be testable in the ground environment, fair statement?

Bob: Yeah, that's correct. That's right.

Interviewer 2: So how do you think. . .

Bob: But the problem with that is the limitations how close can you get to the orbital configuration and still do something on the ground and with the wrap-rib, we could get a lot farther than any other antenna structures we have ever looked at.

Interviewer 2: So, along those lines, this kind of leads into a test as you fly, fly as you test discussion. And these large antennas that you are referring to are almost impossible to do test as you fly because of their size.

Bob: Yeah.

Interviewer 2: Okay, so what kinds of things did you do to work around that to stitch together a V and V [Validate and Verify] program that still was integral and made sense or made you feel good about launching the hardware?

Bob: It was ground based launch, I am sorry, ground based testing of a launch configuration very close to it and so what we did we had tracks on the ceiling, we had cables (I'll tell you about the cables in a minute) every few feet on the rib and those went up to ball bearing rollers on tracks on the ceiling to follow up the exact geometry of the rib as it opened and so what we did is we counterbalanced the cables so we can adjust that. Some of them we even had actuators on so we could change the tension as a function of position and so the idea was they have just enough weight to hold it which was not—in other words—not counting on its stiffness to hold it. So, it being kind of a free state, so when the rib is open, if you pull the rib, pull it down like that, it would just slowly go back.

Interviewer 2: Kind of weight less.

Bob: Yeah, the other thing we did then when we have the rib out, just the rib by itself, so before we put the mesh on it, then we compared the geometry of the parabolic shape at the top of the rib with that of the parabolic shape we measured with a laser when it was in a floatation tank and it was really near free-free.

Interview 2: Excellent, so you were able to replicate. . .

Bob: Yeah.

Interviewer 2: . . .the curvature very well.

Bob: Very close compared to the wavelength because they were typically used at long wavelengths. Ah, what was the wavelength of that, I think it was an L-band antenna.

Interviewer 2: Okay.

Bob: Wavelength something like this [holding hands ~20 cm apart], but we had something most concepts can't have, we got a 0G geometry of the rib in its orbital configuration in that floatation, that was about the minimum load we in-plane load. If it torques a bit it's somewhat forgiving, I can discuss that issue later, but it was very, very close and so then once we got that we put 4 ribs on a hub, the right diameter hub, okay. Once we got that lined up, we took the ribs off and put the tension between them, and because even though it was like 0000000 pounds per inch tension that changed it, so then we put it back on and we adjusted it, so with effects of the tension loading on 4 ribs, only looking at the 2 interior ribs because they didn't have the boundary conditions of the outside ones, so we got close to that. When we repeated that with the mesh on it and got the same very close compared to the wavelength, you know, 50 to 100 to 1.

Interviewer 2: So, let me make sure I understand the geometry right, so the wrap-rib had an inner portion at the hub and an outer portion at the periphery, is that correct or is it all one continuous piece?

Bob: The hub was like this and the ribs came out and the hinge was like this.

Interviewer 2: Yeah, so the hinge was attached at the hub.

Bob: At the hub and the hub had the right geometry to follow the ribs as they continued to cross.

Interviewer 2: If the rib was one continuous composite piece I assume.

Bob: Yeah, it is a single rib coming out, the hinge hinged this way, this was the axis of the hinge.

Interviewer 1: So, it wrapped.

Bob: Yeah.

Interviewer 2: So, when you were talking about the inner part of the ribs, so I am trying to focus in on here on what the critical tolerance is. It sounds like the critical tolerance is the curvature towards the hub. Do I have that right? Did I hear that correctly or no?

Bob: No, because the rib was very deep in the hub, it was very stiff there. The soft part was the part that was..

Interviewer 2: Was way out at the end.

Bob: Way out at the end, so the big problem was the ones out in the end and we had very carefully adjust it with everything so that parabolic shape. Now, some of the fine points. Because there was a tension field in the rib even though it was very, very small, the shape of the floating rib without that tension was slightly different, so we accounted for that mathematically.

Interviewer 2: I see.

Bob: And a large number flown, they were all smaller still can't talk about, I don't know why, no deployment failures. They all deployed completely because once that thing starts to open unless the mesh or something snags it and what they folded it with some soft material, it was between the folds, we didn't take a chance because the mesh, I think it had 10 spaces per inch, it is molybdenum wire, it is 1.2 mil diameter and then it was gold plated, then it was drawn through a die until they got it, and then the knitting was done, at 10 openings per inch with modified ladies stocking machines, stockings, the ones that are used in the large stuff, not the fine stuff, you know, what do you call that, you know.

Interviewer 1: The fishnets.

. . .

Bob: No, I am not trying to get in detail, but you want to know what is involved and all the antennas I worked on, I funded the developments at Harris Corporation, I funded the developments at TRW on the rigid panel sunflower which is a precision deployable antenna and there are all kinds of papers on that, all kinds of papers in the wrap-rib and there are all kinds of papers on the Harris antenna.

Interviewer 2: So, Bob, you have a bibliography right, of stuff you have written and maybe we could append that to this interview; that might be nice.

Bob: Yeah, I think I could dig more, yeah I have got lot of that.

Interviewer 1: Yeah, so just I want to finish up just a thought here on the test as you fly.

Bob: I'll continue, I'll let you talk, but I want to.

Interviewer 1: Oh, okay, sure, keep going.

Bob: I want to get through one concept which is the most challenging and it was the easiest one to compensate for in terms of reliability because we not only did this, but all little guys, all smack wide open and perfect and you get a combination of large numbers and smaller ones on the ground. And this approach at the time it was done, I think it was done in the 80s, done a long time ago was, I did a lot of consulting for the military, it is not on the books, it is not in the papers, but I was allowed to do that. Why did Caltech allow me to do it because they are interested in the technology for big antennas and they can't afford to develop them and the really big stuff like that 55 meter, NASA could barely afford that. And the worse, the big one of course was the ISAT antenna, they had four groups, I have a paper on that one, four groups of manufacturers, they were all competing, they were competing Astros competing with theirs, we had L'Garde competing with theirs and of course Harris was in there and that was supposed to be functionally—this was in the paper, I even have copies, extra copies of that paper.

Interviewer 2: I think I have that paper, Bob for the benefit of the audience, the ISAT was 100 meter deployable, is that correct?

Bob: The ISAT?

Interviewer 2: Yeah.

Bob: ISAT functional was 300 meters.

Interviewer 2: 300 meters and was L-band?

Bob: It was a phased array and I don't remember what the input. They had, they could adjust the amplitude and phase of all these elements and they had in the hard panels, so it is hard to say what the enemy of precision was knowing what it was in orbit and knowing how much their adaptive system could compensate. Now, you may not want to include this, but the problem I have run into on all the antennas that are adjustable, is electrical people, say "oh yeah, you know, you guys don't have to kill yourself. We know the higher the precision, the heavier it is going to be, so you can do meaningful ground test. We know all that and we can do all this compensation." And it turns out in my experience that approach is total bunk because they don't even know what they can compensate for in these big structures electrically even though their models are very elegant, until they actually get some, some real hardware and we could get some idea if it deploys successfully, that's not enough. What is the initial precision that is the second large one, deploy it and you have to know what it is because the super-duper-ooper unbelievable sensing and adjustment systems are very limited and the more they correct, the lower the efficiency of the electrical performance goes. See they are all tied together.

Interviewer 1: Now, this active compensation is this electrical in the backend or this is actual shape control that you are...

Bob: No, no, and these big ones were not proposing shape control because there are too many pieces.

Interviewer 1: Okay, yeah.

Bob: There are hundreds if not a thousand pieces.

Interviewer 1: Okay.

Bob: And that's why the kinematic deployment reliability is going to be so low. I wouldn't want to be someone to tell General Powell that I think this thing is going to work. And the best is... you know the most complicated one—this is a fact and the people won't like that—the most complicated one in the late-... especially in the ISAT—was the Harris antenna. They had lanyards and they had pulleys and they had hinges and they had latches and they made some prototype hardware and it was the finest kinematic hardware I have ever seen, the quality of it. It was better than some flight stuff I have seen the military fly, it's beautiful. I don't know of anyone, makes in hardware I have seen—really seen not just took like a quick look—seen it with the pieces together, seeing them line it up and seeing doing deployment. It probably is the best in the world, but the problem, the part count is astronomical and that's what kills estimates of reliability without valid a test result.

Interviewer 2: So, Bob, for background, ISAT made it up the PDR, right before it got cancelled, do I remember that correctly?

Bob: Yeah.

Interviewer 2: Okay, so at least up to that phase of the design, what would you say are your major lessons learned in that development activity?

Bob: Which concept?

Interviewer 2: Well, you had your competing contractors, right?

Bob: Yeah, and there were groups of contractors.

Interviewer 2: Right, right, so let's restrict it to technical lessons learned as opposed to programmatic lessons learned because programmatically it was extremely complicated, just your technical lessons learned based on trying to design a 300-meter long antenna, what would you say the big takeaways were from that work?

Bob: Well, basic concepts were process, Astro had a very fine process, probably some of the best work and it was mechanical, it was a takeoff on some of the, what it was, remember the AstroMesh.

Interviewer 2: Oh, yeah.

Bob: That had a ring around it.

Interviewer 2: Sure, the perimeter truss.

Bob: Yeah, of course, don't laugh, they opened that truss up and made it straight and some modifications and they built a pretty big model, I think that my papers says what it is. Boeing even at the last minute with the contractor group that was working with L'Garde and they had an inflatable deployable rigidizable truss because all of them had rigid panels on the top, of different sizes and they had to have them because of all the transmit and receiving modules on. It was a true phased array antenna and so the big concept there was how do we open up, how do we hinge it and what kind of a truss... we are going to have to have some kind of a truss. You can't have these barn doors hinged and flapping around in space, how are we going to do it, we got to have some kind of truss. You got to have depth.

Interviewer 2: You needed the rigidity.

Bob: Absolutely.

Interviewer 2: Yeah, so that's a lesson learned. You needed to have rigidity to control deflection.

Bob: Yeah.

Interviewer 2: Okay, did you see any advantages between the inflatable truss and/or the mechanical truss, compare and contrast the concepts?

Bob: The biggest difference was in the cost. Inflatable truss was probably six or eight hundred times cheaper than the mechanical truss and there is orders, orders of magnitude difference between the potential, the part count and potential failure modes because inflatable is soft and it folds up in a nice smooth container and no matter how hard the booster shakes it, you're really not going to crack it or do anything. So you are not designing that for launch. That's the big difference between all of them, that's why I flew the IAE [Inflatable Antenna Experiment] and a container about 3 times the size of this table. They don't break in launch. They are flattened and they are folded up. So then, but the disadvantage of those because they are inflatable and you do all you can on the ground, the disadvantage of that class of antennas is large variations of initial position, that is the bad one because they are inflatable and they do stretch a little bit and if you do even fold it up for a long time, there is probably some strain relief in some of the load-carrying fibers, different fibers are used, Kevlar, glass, depends on whose concept, combinations. They actually use combination of fibers, with a positive CTE, others are negative CTE, and balance the design so they are thermally stable at the orbital temperature.

Interviewer 2: Okay.

Bob: That's another big advantage of these inflatables, so from reliability point of view and the cost point of view, if you can get them to work, that's the ones that I think have most potential in these gargantuan sizes.

Interviewer 2: Did you ever get to the point with the ISAT program where you can do a performance based comparison between the inflatable truss and mechanical truss for the ISAT configuration?

Bob: We just started; we started at the preliminary design review when the faucet of money was turned on.

Interviewer 2: I see. Was the inflatable looking pretty promising?

Bob: You may not want this, probably don't want to publish this. Only after it was taken over by another vendor.

Interviewer 2: Okay.

Bob: The other contractors that I won't mention didn't have the faintest idea what the inflatable could do and they were buying a lot of bunk from vendors about what they can do.

Interviewer 2: Yeah.

Bob: You know some vendors are always "oh, just tell us what you need".

Interviewer 1: Yeah.

Bob: So, this is that part is sensitive. And so one of the big meetings, contractor meetings with DARPA, one of the only unpleasant jobs I had, after the four groups got up looking at me, namely me, I had to get up and tell them what was wrong, what they said that we don't believe, I didn't want to do that and on the concept a vendor was building, I had to tell them, I said this isn't working, so they said, is there anyone that could make that work and I said a certain group does the best structural systems work I have seen, I have got a number of contracts with them compared to anyone in the country, compared to JPL, compared to the other centers, the other states, it was the best in my opinion and of course JPL you know they have this and I never published that but they were the best, clearly, that's it.

Interviewer 2: So, Bob, the conversation is leading towards programmatic lessons learned now and again as we said earlier...

Bob: No, I don't want to go into programmatic.

Interviewer 2: What's that?

Bob: I am not interested in going.

Interviewer 2: Okay, that's fine with me. I thought maybe you had some pearls of wisdom in a general sense about managing these things.

Bob: But you asked a question, I had a long answer. How do you compare them? You can't say one vendor was the best concept. It's only the best concept when they were doing it for another vendor and they built the hardware with another vendor's help and some of the mechanical parts that weren't good, the vendor built their own and assembled it in a big scale and they were in the process of testing it. So I can't just say well "this concept was it", it's tied with the programmatic.

Interviewer 2: Sure.

Interviewer 2: Did you find back to the technical for a moment on the inflatable. Of course there needs to be an infrastructure for doing the inflation, was that any kind of a driver?

Bob: Inflation system?

Interviewer 2: Yeah, was that pretty manageable?

Bob: We don't know. We had, because on the experiment, the IAEs. The inflation system the struts were perfect, but the inflation system with the lenticular inflated, but not, was not as high as we want it, and we believe there are some problems in some of the flexible tubes and the control involving the technical tubes, so it was a complexity of the tubes, the way they packaged the tubes and whether it was the absolute control which it should not have been, we don't really know because we finished the experiment, we kicked that out and they firmed up.

Bob: Okay. I have a big umbrella in case someone reads this and says "I'm going to sue Freeland".

Interviewer 2: I am going to take a pause here for a moment, Case, I think.

Bob: Jan has submarines for lunch!

Interviewer 2: Picking back up after a break on the TRW sunflower.

Bob: TRW sunflower was very clever and I have got some write-ups, so I can give you pictures of it and I have also got reports from TRW. It was a high precision rigid element deployable antenna supposed to go at much higher frequencies than any of the others. There was a bunch doubly curved high precision rigid elements and they had hinges and rods and the problem with that, I think the biggest one they ever made was something like 6 or 7 feet, not meters, and the DOD gave me some money and I gave them a contract to do a study to determine how high they think that could go. Now that was long, that was back in early 80s, I will have to go back and look at the papers. The challenge, the biggest drawback, was weight and mechanical packaging efficiency. The efficiency wasn't very good compared to the size of the elements; it was maybe three times better, four times at the most better than the size of the hard elements. They were done with high precision graphite-epoxy honeycomb and you could make those in small sizes, very high. So, then, what is the challenge? The challenge is coming up with hinges, mechanisms for a given stowed diameter, how can the diameter can you come with, anything you can back it, how can you align them up. So, total precision is going to be way less than the precision of each individual doubly curved panel. Those are very high precision, so whole thing is, got you lining up relative to the edges coming together like this or gaps and the tip in different directions, those alignments, those are, they animate. So what they are trying to do is come up with a crude concept where they could do a better mechanical packaging ratio than 3 or 4 that's stowed-to-deploy because something like 6 or 8, then the mechanism that would enable that to fit together is one near-continuous surface, then to go from there and work on that mechanism to provide that to get the proper alignment and tilt relative from panel to panel—all of its neighbors. That is really pretty elegant because there is a lot of military applications that they want to go up to 15 to 18 Gigahertz.

Interviewer 2: Oh wow, okay.

Bob: And they want, and they wanted something like 10 meters and that's what we are shooting for and the guy did it all, the genius on that, a friend of mine, Dr. John Archer, really a good man, known him for years, a nice guy, just really good, and in fact I am 85, I don't think that he is here now.

Interviewer 2: So, Bob, what were the lessons learned then on that particular design on that petal design, what's critical if you are going to go this doubly curved rigid surface?

Bob: The biggest, and oh, because it was solid element and had all these other elements, a big feature of it, you could put a... test a big one on the ground because the mechanism support structure had to be pretty hefty. The biggest challenge by an order of magnitude was to develop the mechanism to fold it up in ratios of something like 6, 7, 8 to 1. It could be done; you had to be very clever. What they did do, they did do enough to come up with some rough approaches, what is the basic fold up configuration. There is a basic one; they are not really nicely stacked. The highest packaging efficiency would be they would take each doubly curved element and stack them up like saucers—that's the maximum you could achieve. So, starting with a stack of the stacked elements, what would you have to do to move them out and line them up and that is what they were working on and I think they doubled, they got close to tripling the basic one, but it was a concept that was not well funded. I think all they got in the early 80s a few hundred thousands, six or eight hundred thousand, I think we could have made it!

Interviewer 1: You have seen this guy, the flasher, the little origami inspired.

Bob: I have seen that. I like it.

Interviewer 1: Yeah.

Bob: I like it.

Interviewer 2: Is it the Brian Trease's design?

Interviewer 1: Its Trease, Pellegrino, it's Starshade. There is a lot of people working on this. And so each facet there could be your own doubly curved... the fold lines define your mountains and valleys and then you can...

Bob: I told you I wrote Caltech about it.

Interviewer 1: Yeah.

Interviewer 1: That's exactly what.

Bob: Just this. You see he has got a lot of good stuff and the value of, the integral value, is gigantic!

Interviewer 2: Yeah, for the record here, we are looking at a radially-folded piece of origami very intricate pattern, the JPL flasher design.

Bob: Actually I think it's one of the cleverest ones around.

Interviewer 1: I carry this with me at all times just in case I am going to talk to somebody about it.

Bob: That's elegant, you come up with that, no one is going to point to you and laugh.

Interviewer 1: It's... we were thinking about it for solar panels which don't have any real flatness requirements, but you want your real estate, you want your aperture.

Bob: And that well it needs a few degrees.

Interviewer 1: Yeah, and so if each one of those is a thin film solar ray, you can get very good packing efficiency and this goes to a nice convenient cylinder, all your launch shrouds are cylinders, so this has a—here you go, we pull backwards, I think it is good.

Bob: And the astro truss, Mark designed, he and my late friend, the late Dr. John Hedgepeth came up with some of the truss design. He came up with that he was going to use to support the panels in the ISAT program.

Bob: Really clever, awfully good work.

Bob: Going all the way back to the Astro, that was done by Bob Crawford and John Hedgepeth, that wasn't just John Hedgepeth . . . one of the original one was Robert Crawford, he is gone now. The reason these are dirty is because I was up working cleaning the boat this morning.

Interviewer 2: Wouldn't even notice Bob if you hadn't mentioned it.

Bob: What?

Interviewer 2: Wouldn't even have noticed.

Bob: Jan cares, I don't. So what other concepts and I said it depends on the concept and the precision deployable because almost by definition of the original elements you are going to have to have to align that up and support it, it probably is going to lend itself to larger size ground testing. That is qualitative, not quantitative.

Interviewer 2: Yeah, so Bob, that's the theme I would like to return to you, this goes back to the test as you fly stuff. The test as you fly, fly as you test, right. At some point, you break the bank. You can't do test as you fly. It's just too big.

Bob: Too expensive.

Interviewer 2: Too big, too expensive, right, so lessons, advice when you get to that point for example let's say you are building a, I am going to make something up here, you are building a 50-meter circular antenna and there is just no mechanical infrastructure that you can get a hold on that will allow to do a deploy on the ground, so how would you approach that?

Bob: You said you can't test it in the ground?

Interviewer 2: Yeah, not the whole thing, maybe pieces.

Bob: Up to what size can you test it in the ground?

Interviewer 2: I am going to make something up, just so it forces you to think about a different paradigm. Let's say the I think I used what 50 meters, let's say that your facility will only support testing up to 25 meters, let's just say, so what do you do, how do you attack that?

Bob: That's not, that's not that much of a challenge, that's fairly easy to do.

Interviewer 1: Alright.

Bob: So, the biggest one you can test is 10 meters.

Interviewer 2: Okay, fine. Okay, make it.

Bob: Now it becomes very, very interesting.

Interviewer 2: Alright, let's talk about that.

Bob: So for a given size, launch vehicle and its allocation of mass, if you look at those, you think, what, what's got the highest probability of being able to have a meaningful ground test, tricks you play on the ground, so you can present by a factor of 5 to 1, 2 to 1 depending on concept, is doable on some antennas probably, a simple one like a Lockheed wrap-rib, that's pretty good, but what will I do 5 to 1, no way to ground test it. That would force me to look at concepts that would fold up and have that packaging ratio. They would fit into the canister of the launch vehicle and I will take a rough look at the weight to see how that compared to the allocation for that. That's what I will start with. I just wouldn't try to make a guess and of course the important thing is the operating frequency and the associated precision, is it a reflector antenna or is it a phased array antenna. If it is a phased array antenna, is it a passive or an active phased array. One of them is just passive, it gathers up electromagnetic energy and processes it with a bunch of multiple seekers. If it is active, it goes both ways, you know that. So, those are some of the things I think about, then I would look at all the, copies of everything up to 10 years ago after I left, I was in other work for the DOD, it was not antennas, I was doing work for other agencies that I will you about when recorder is turned off.

Interviewer 2: So, back to the testing thing from.

Bob: Is it off?

Interviewer 1: No, it is on, I was just making sure it was on and then we'll finish.

Bob: Okay, when we finish, I will tell you.

Interviewer 1: Yeah, yeah.

Interviewer 2: So, back to the testing thing then Bob, so how do you stitch together a test program if you are facility challenged, what do you have to do?

Bob: Lockheed wrap-rib was facility challenged, so wasn't the Goodyear inflatable work, it's covered and one of my publications, it was done for the black Air Force. I ended up according to some knowledgeable people. They only set of declassified photos which I still have and they were declassified, so I can use them put them in one of my... and they built some big stuff. You have seen them one of my reports?

Interviewer 2: Yes, I have actually.

Bob: Good and that, they built some pretty big stuff. The fact that they even built bigger than the stuff I have pictures of.

Interviewer 2: Oh, really.

Bob: So, the facility is challenged there, so for things like those, a rigid, something in the base, rigidizable stuff, for instance, if the concept—don't laugh—is simple enough, truss, it's like a truss, it holds some stuff on it and you want to know, can you come up with a truss to support a 50-foot aperture of some kind. There are truss designs you could come up with. You couldn't build a whole thing, but the important part is the deployable part. You could build elements of a 50-foot truss if it's simple for all you have is a bunch of tools, that's easy to do, so it depends again on what the concept is and I would, if I couldn't test a big one, in addition to the things I gave you, I would say, well, what concept or combinations of concepts, can use a bunch of simple structures, so I can make the things, full scale and not have a big facility to make a them work. And so I can demonstrate manufacturing capability for large concept elements, and I can demonstrate packaging techniques with a much smaller one, and I can demonstrate deployment potential by taking certain elements or subsystems of the support truss and demonstrate those in such a manner that makes sense, that the integral of all of these demonstrations of actual hardware plus the computer models might give you enough confidence that you might want to take a chance and build one. Now how do you like that answer.

Interviewer 2: That's a great answer.

Interviewer 1: I like that, can I poke at that for a second.

Bob: You can focus on anything including the parrot.

Interviewer 1: I am thinking of, we have got a 50 meter we want to do, we can build a 10 meter one at pretty high fidelity but its one-fifth scale.

Bob: Right.

Interviewer 1: And then we can build one hinge at full scale, but we can't put the rest of the antenna on it because we don't have the room, just do a little cookie cutter focus in on one hinge at full scale, one concept at one fifth scale, those two plus a mathematical model?

Bob: That's what I said, because because those can be looked at as subsystems, subsystem assemblies, not the whole antenna. If you can demonstrate the subsystem assemblies that can be used and all these things I described, yeah, that's what I would do and I would keep all this in mind as I look at concepts, I will try to get smart people, I am not smart, I don't come up with clever things. I can take a look at subsystems in clever stuff and sometimes I can see how it goes together, but when you have done it a few times, you will get because I am not a designer. I think in systems, that's why my title is "structural space" or "structural systems" engineer, that's what I am.

Interviewer 2: No, the previous answer I thought was great, I mean, you are really the term that I like using stitching together the elements so that it is to use a mathematical term that it is piecewise continuous.

Bob: Right.

Interviewer 1: So it hangs together from end-to-end.

Bob: That's a different way of saying it.

Interviewer 2: Yeah, yeah, that's exactly what you said. It requires a lot of engineering judgment, right, to do that well and that's maybe the hardest part of it all.

Bob: You know I didn't think of it, but you are right. That's the whole problem

Interviewer 2: So yeah.

Interviewer 1: If I was going to do one. . .

Bob: Like what?

Interviewer 1: If I was just going to do one test, which one do I do, subscale for the full system or the most critical piece at full scale?

Bob: You start with the most critical elements in subsystems, like if you have to have high precision, exactly.

Bob: Subsystem.

Bob: High precision initial geometry and start with that.

Interviewer 1: Okay.

Bob: Because if that isn't good well you assemble a piece of crap together, the whole antenna is a piece of crap you know.

Interviewer 1: Sure.

Bob: Well you've got the right idea. What's really the heart of it, what subsystem or element, or maybe some manufacturing in shape of something that's really complicated has to be really light, really stiff, really thermally stable, really high manufactured tolerances, something like that, that's about as bad as requirements which you could get to make it and especially if all this is needed in a big piece, so if the concept is dependent on that, it could be, don't laugh, if the concept is simple enough, it might be just manufacturing these really hot elements that are hard to produce, exotic materials never been used before, great surface finish, zero coefficient of thermal expansion—at the operating temperature on orbit—all of these things which drives you to some kind of elegant composite, yes absolutely.

Interviewer 2: So, Bob, I want to poke at scaling a little bit.

Bob: I was hoping we would avoid that one.

Interviewer 2: So. Back to being constrained, let's suppose you build a 1 to 3 scale model of the thing to demonstrate kinematic reliability, what is your experience on how well subscale models scale up or vice versa for that matter, well are there some "gotchas" there?

Bob: The only one I have experienced with is the Lockheed wrap-rib, Harris has got some data there, but I decided not to get it. They would give it to me because they know I didn't give secrets away. I can give opinions about what their product will do, but I am not about to give away anything they do, anyone, but yes, with Lockheed, let me only think, I haven't thought of that for a long time, on the ATS-6 (I wouldn't use the word ATS-6, I would use the diameter of that one) and they scale that up. They never built one, the biggest one they built is they built for me for NASA. I asked them how good their models are and they said for one with a lot of kinematics, no more than 2 to 3 to 1 and for very simple one like theirs where there was just one hinge, they said on that one, they felt good about going up, they go up in a factor, they said 8 or 10.

Interviewer 2: No kidding.

Bob: Yeah.

Interviewer 2: Oh wow.

Bob: Because it was simple.

Interviewer 2: Okay.

Bob: And had one mechanism.

Interviewer 2: Okay.

Bob: And they folded the mesh, I described for you in detail. In each fold, they had some soft material, so they opened up, it would go off in space, and there was nothing to stick or catch.

Interviewer 1: Was that like a little piece of foam.

Bob: Yeah, really light stuff and stuff that was rough like foam, so it wouldn't slide around. The other nice thing about the Lockheed concept on the large side, the mesh was so fine that if they did catch and ripped a piece out on a large aperture it really wouldn't do that much to degrade the performance unless you had high side lobes. Small errors on reflector antenna surfaces will screw up the electrical transmission by throwing energy into the side lobes in the main beam. That's the way they work, so even though it might be a bump like this, the size of a potato in an aperture that is bigger than this room, the DOD doesn't want energy in the side lobes for detection and a radiometer is placed on a parabolic reflector antenna, you want over 90% on-axis gain performance. So what does that do. That makes the deployable antennas which are all flats approximations of something like the Harris antenna has all these tie points, it's near, it's a good way to go, but it's not doubly curved. So a whole bunch of little diamond shaped flats because between the ties, they are flat, and the Lockheed it is between... so they are all flats approximation. You have an antenna with a surface tensed at one area kind of a big bump or a hole or misalignment of panels, that will throw energy in side lobes. It doesn't take much energy in side lobes to knock the crap out of the on-axis gain on a radiometer application. Now the military looking at bad guy stuff and the science looking at icebergs, oil spills, all those temples they found in Central America and are 100 foot of growth, they didn't even know who is there using a long wave.

Interviewer 2: Right.

Bob: Radiometers. The roads in Egypt that the old chariots used to go on that were tens of feet under the sand, who do you think located them? Long wavelength radiometers!

Interviewer 2: Right.

Bob: You know that.

Interviewer 2: Yeah. The lost city of Ubar, I remember that. That was one of the JPL SAR radars actually.

Bob: Okay.

Interviewer 2: Yeah they did a NOVA on it.

Interviewer 2: Cool. We take a break maybe Case. [Break]

Interviewer 1: So, recording again after lunch. One of the potential uses for this handbook is somebody has got a problem and they want to know how to approach it. So, when we were talking about doing a subscale full model or a full-scale component, not that there is one answer but how do you even start the question, how do you decide what's your most important aspect of the problem. I think that sort of question would be in the minds of... whoever's reading this book is going to have maybe something in mind where they want to figure out how to proceed.

Bob: Well that's experience, I'll tell you precisely what it is. You have to be, familiar with essentially most of the concepts, not every one or technology level must use a scale 1 to 16, one's a piece of paper with some drawings or some numbers and 16 is flight return data. If it's 2, 3, 4, something like that, you can't know all of those. They have to have an understanding like I do or did, of what the level of technology is for this general class of structures, in particular the class of structures of which you want something in that class they want working on orbit. It's something in that class you know, what the booster is, you know what the hard-line limitations are, you have to be familiar enough with the concepts around like for instance—let's see how do I put this without mentioning names again—the "highly complicated kinematic concepts"...

Bob: ...doesn't give me much flexibility. The company is... they're just too complicated, the part count, it's just a—it doesn't lend itself to taking a particular member or subsystem out if the whole thing, it's all mechanisms and lanyards and latches and locks, and actuators, and all that because if the whole thing is covered with those, that really makes it difficult to go through the process that I'm recommending. I'm going to look for a concept that you might improve the mechanical... I don't expect to pick one and then have it fit; we'll probably have to modify it. But that stuff, a lot of smart people around, well there used to be, when there was interest and the interest is all gone now, ever since all the government funding for the good stuff went away!

Interviewer 1: So, Bob as usual in design, simplicity is king.

Bob: Yeah.

Interviewer 1: No exception in this field.

Bob: That's right. So, you can look around and look at different concepts. Example, if you have a long thin cylinder and you know the concept, you know in terms of packaging configuration what might fit there, or if you have a short fat one, which is a tougher one, that leads you to another concept. Short fat concept: perfect for inflatables, because you can, you can fold them any way. You can fold them up, you don't want a tiny cylinder a long cylinder, if you have fat cylinder, you could store them in that... In their concept of packaging you might have part of it in a short fat round container, and then maybe certain long truss elements might be deflated and folded up and put along the cylinder, these are just examples.

Interviewer 2: So, I have question (Case maybe on a slightly different vein). So Bob you have got just a tremendous amount of experience over 50 years worth of experience in this area.

Bob: How many years?

Interviewer 2: Over 50.

Bob: Over 50.

Interviewer 2: Is that right?

Bob: That's right. Over 50.

Interviewer 2: And so as they say "there's nothing new under the sun" and you've been looking at a lot of concepts that are now starting to regain some currency. Where do you see the field, the deployables field moving at this point in the game?

Bob: The only thing I know, is what the DOD wants. They all call me up. Seemed to work close enough to what they wanted, I seem to be friends with most of them. I still get calls. The last person that I talked, spent a lot of time with, for science: the world famous guy, Doctor, Professor, Bernard Burke from MIT. And he headed up the large science instruments the telescopes and so on, for 3 of the 16 science committees that advised NASA at that time the best known being the national science foundation, national academy of science, all of these. And I really liked him, and he seemed to tolerate me, and we were paying, oh I'm pretty sure he's gone now because he was oh, 12–15 years older than me, they call up, they said, you know I'm interested in the direct detection of planets around foreign suns, that's 12 or 15 orders of magnitude difference in intensity from the star or reflected light, so we wanted to know, and the big monolith, we might take to space, how big could we make them in diameter, optical quality. And then the answer to that was clearly there would be segmented elements of ultra-low expansion glass (ULE), that's what it would be and because of the boost environment you might look in to the technology for ULE glass elements that incorporated flexible fibers in North Florida who was doing a lot of that with DOD money, had a $2M contract with them before it got cancelled for programmatic reasons, United Technology Research Center in Florida. And this question was so long ago, I said "what drives you in that direction?" The answer is very simple: EVA! You take the telescopes and stuff on orbit and you have EVA assemble them, and then you drag them to the—the operational orbit. See, the initial space station concept was under development at NASA Langley research center, where my good friend Richard Russell, the famous senator Russell one of his relatives, still talk to him! And he headed up the study. The initial configuration had an outside assembly port, for orbital telescopes and antennas, even deployables which they could take up and deploy, if they didn't deploy all the way, they astronaut could, go out there and kick them in the ass or make some adjustments and get the operational configuration before you are taken to the application orbit. And then you could do all sorts of good stuff, but hey, who in headquarters dumped that, that was the first big element that they dumped and that was the dumbest thing and the dumbest decision on them, that's how all these agencies for science or VLBI, 50 meter, 100 meter diameter reflectors on orbit in conjunction with the DSN and ESA/ESOC their system of ground antennas for VLBI. The snag was one of the big ones for radiometry the DOD ops. you know what they wanted it for, deep space comm, send a spacecraft out with a decent antenna for a change, so when it folds up and then goes out, 12 or 15 feet, doable? Yes! Very doable, so anyhow I'm just giving you one example, there's lots of examples.

Interviewer 2: Yeah, so ideas regarding on orbit assembly, in-space manufacturing, which have been around for a while are actually starting to get popular again, particularly with the the...

Bob: Trusses, they're in my papers, a number of trusses were assembled on orbit with EVA. You have got to go back into the 80s ,that is all documented. Most of the concepts were done at NASA Langley Research Center and the ground based assembly—this is another thing that you're going to like.

Simulating the orbit in the zero-g tank the water tank at JSC, they have got that big tank, that must be, 2-3 times the size of this room.

Interviewer 2: I'd love to take a swim in that sometime.

Bob: I did.

Interviewer 2: Oh did you?

Bob: I was a certified aqualung diver. Went down there in the 80s, I said, "sure be good to go in there" and they said "you certified?" and I said "you bet!". "The equipment's in that room, and they have all size suits, even for skinny [guys] like you."

Interviewer 2: Nice, I would love to do that. I would love doing that.

Bob: So that for certain types of applications, never thought I was going to bring up anyway, if you get a master job, of buoyancy tricks, you could take certain types of structures, the gravity is still there but you could do a lot more and you could have in that tank. There are certain subsystems and elements, you could play games with in that tank, that is another simulation capability that most people have forgotten—EVA crapped out, crapped out for political reasons because they... technically they were, I mean, they had... someone had to have their head up their ass when they made that decision. And you can leave that in if you like. As my quote.

Interviewer 1: Do you think that's still a winning strategy to chase, is to include EVA?

Bob: Yeah, for orbit? Oh, yeah.

Interviewer 1: I'm just thinking of these big mesh reflectors, with the perimeter trusses that deploy out and the original concept was that you deployed them outside the space station so that somebody can see if they are correct and then "tugboat" them to the preferred orbit?

Bob: Yeah, that's the—that's a capital idea. That's absolutely perfect. Because they can do that without a lot of risk, there's a lot of things they could do if something sticks, even absolutely destroyed the concern about reliability. Problem is towing it, because getting it from there, from a few hundred miles, to 23,400 miles.

Interviewer 2: That's another story.

Interviewer 1: That's lot of deltaV and once it's deployed it's a little flimsy, and so you can get in trouble I would imagine.

Bob: That's right, that's right. With any kind of thing except a very moderate thrust, yes. Hadn't thought of that for years until you guys came out. It's obvious. Anyone want some more tea?

Interviewer 2: I'm good thanks, Bob.

Bob: Jan, do we have any kind of pop in the house that has sugar in it instead of this diet stuff? We have a young man here and he would drink if it were here.

Interviewer 1: I found one for lunch, thank you very much. I got a little green one, and that was just perfect.

Bob: We usually try to keep it, because you always have some really nice fat friends.

Interviewer 2: Anything else come to mind, Case, huh, I'm kind of tapped out actually.

Interviewer 2: Let's see, well personally I have hired a few Ph.D.s into my group and we are working on space structures technology, what – some of them folding origami stuff, some of them solving stability for precision booms and trusses. The facility is a challenge, what has the stability to be useful. The metrology is a challenge, what allows us to measure what we care about. . . the mechanical, thermal, acoustic air, there's all these challenges that I have got that I am chasing, but I imagine you chased down these roads already, some dozens of times. What should I share with the new members of my team to get them started on the right foot.

Bob: I'll get your card, let me think about that. You know that requires some thinking. That's not right off the top of my head because I haven't thought about this for 10 years. When my DOD friends call, they want, they want to talk about ISAT-type structures, they want to know of the concepts were any of them worthwhile, and why. That's why I have a huge umbrella. Because, I tell them the truth. I tell them which, remember one thing, you want to know how big to reach this class of applications that can be developed with capability for evaluations that I'm supposed to think of, that are worth being invested in, that is the way they approach me. If I were in their shoes that's exactly the way I would approach it. Because most of these people they're so much smarter than the NASA people, headquarters people, these guys are really smart. The best example was Joe Gersey. Doctor, Professor, Joe Gersey now CEO Joe Gersey. He's one of my favorite people, he's smart, he's tough, and he's really funny. You guys would really work great together.

Interviewer 1: Yeah, I met him, out at your structures conference you guys had out here at the Embassy Suites, I think he made an appearance, remember that, it was probably about 10 or 12 years ago, you had John Egan out here as well, those were the big names as I recall. Oh, Zapman was out here too. Yeah, yeah, poor old Zapman.

Bob: He was a nice person and I liked him, and he was very very honest. He was MIT top graduate, and his electronics, the feeds and things, he was one of the ones that kept pushing the electrical connections to take all the big problems away from us, and every time the question came up, "what do you do with electric performance for making all these adjustments, of this magnitude?" well, I'm still waiting for the answer.

Interviewer 2: Was he an electrical engineer?

Bob: Yeah.

Interview 2: Okay, well there you have it.

Bob: He is very good—no he is really—he is literally—he is a good man. He is a nice person, he has been out here to dinner a few times I like him. The fact that he really did not think, or have any respect for our discipline, it was okay, because he didn't know. He's a nice person. He wrote a nice thing that I will need to frame and hang up, thanking us on the work that I did as I was the chief structural systems technology on the program. And typically they don't farm that out and I did that for them wearing a DARPA hat. That was—that was very rare.

Interviewer 1: This is wearing a DARPA hat in your office at JPL?

Bob: Yeah, on that program, only on that program. I had to wear a NASA hat on the LSST Large Space System Technology program back in the 70s in fact I went back to Langley for a year...Yeah 78. In the 70s... No, no, no, no, Jan...what year did I go back to Langley for a year?

Bob: I wore a NASA hat for that program, few hundred million, and the ASEB was held at Stanford, one of the big guys was doctor/professor Holt Ashley, world famous in control dynamics expert. And so there's a big committee up there and, wearing a NASA hat for Langley, I got up and personally defended the program, technically defended the program. They showed these different concepts in different areas and I had to address those different concepts, and use that to structure the elements of the program. Yeah I did that by myself, I didn't get any help.

These old icons like Tom Cambpell. Headed the radio frequency and did the feed systems and my friend Dick Russell who did the system studies. We all worked together. Especially Bill Walye, who is the systems guy back there, they had me give the pitch. I have got evidence to show that. Now, the different contractors like for the Sunflower at TRW in the other programs and for the people come up and talk about the cost and what they could do for the cost, I can't do that. So, yes I did wear the hat. Did I like that, yes I liked that. The more hats the better. No, I like different things, I'm not hung up on any concept for anyone's work. Hung up on good people, yes, because, you know there's a shortage of those… makes such a difference, doesn't matter what they do. Okay, go ahead, beat me up again.

Interviewer 1: I'm just digesting everything that we have chatted about today, I've been taking notes I am going to think over the questions here, what else should we cover to do our due diligence?

Interviewer 2: I think we've done a good job. I think what we need to do is get these things transcribed and we have to resort them, because we moved around a lot.

Interviewer 2: So, I guess we'll wrap this up.

APPENDIX 3

Written Interview with Dr. Martin Mikulas
(Courtesy of Dr. Richard Papa, NASA LaRC)

BIOGRAPHY FOR DR. MARTIN MIKULAS

Martin Mikulas is professor emeritus of the University of Colorado and a member of the prestigious National Academy of Engineering (NAE). Mikulas joined the University of Colorado faculty in 1990, where he taught aerospace engineering. Prior to that, he held several positions at NASA, including as head of the space station structures design team, structural concepts branch, and the structural mechanics branch. He holds nine patents and is a fellow of the American Institute of Aeronautics and Astronautics. He earned bachelor's and master's degrees and a Ph.D. in engineering mechanics from Virginia Tech.

INTRODUCTION

The U.S. put its first satellite, The Explorer I, into orbit in 1958. Because of operational experiences in World War II, world-wide communications was a top priority. Due to the low electrical power available in space at that time, high consideration was given to passive communications and the ECHO satellite program was initiated. The challenges of fabricating, packaging, and successfully deploying the required large diameter reflective surfaces were immense. For several years, large resources were applied to the development of membrane technology for such missions. However, the concept of passive communication satellites came to an abrupt end with the rapid development of solar cell technology that enabled adequate electrical power for continuous active communication satellites.

EARLY WRINKLED MEMBRANE THEORY AND VALIDATION EXPERIMENTS

I came to work at NASA Langley in 1961 during the relatively short heyday of the development of membrane structures technology. The first task that I was given was to experimentally determine the validity of the wrinkled membrane theory developed by Stein and Hedgepeth in Reference 1. A decision was made to attempt to experimentally replicate the two example problems presented in Reference 1. The first was the pure bending of a pressurized membrane cylinder reported in the 1962 paper, Reference 2, while the second was the rotation of a hub attached to a stretched circular membrane reported in the 1964 paper, Reference 3. In both cases, the resultant deformations were to be measured during the pre-wrinkled state and well into the post wrinkled state. These experiment were to be designed to match the boundary conditions of the two problems analyzed in Reference 1. However, it was determined that the infinite extent boundary condition assumption of the hub rotation problem of Reference 1 would be difficult to experimentally simulate accurately. Thus, it was decided to solve the problem of a hub attached to circular membrane of finite extent. This problem was solved and results are reported in Reference 3.

As an interesting side note, during the early 1960's, it was common for the branch heads to have world-renowned experts come to Langley to interact with researchers in areas of high current interest. In the structures area, my branch head Bob Leonard had visiting experts such as Eric Reissner, Bernie Budiansky, Lyle Sanders, Nicholas Hoff, John Hedgepeth, Paul Seide, and others come to Langley for mutual discussions of on-going research. As I was presenting my work on the hub rotation problem at one of these informal meetings with Professor Reissner, he asked if I was familiar with his similar work published in 1938, Reference 4. Of course I was not. Upon investigating his work, it was determined that although he used a completely different analytical approach to that of Stein and Hedgepeth, the resulting solutions were identical. Reissner assumed that the modulus of the material perpendicular to the wrinkles was zero, commonly called tension-field theory, while Stein and Hedgepeth assumed a variable Poisson's ratio to account for material gathering in the wrinkled region. Reissner's work was restricted to unstretched membranes for which wrinkling encompassed the entire membrane. The variable Poisson's ratio approach of Stein and Hedgepeth seemed to be intuitively more applicable to membrane problems that involved both wrinkled as well as unwrinkled states. At the time that this work at Langley was being conducted, we were not aware of the General Electric work by Comer and Levy in Reference 5.

DETERMINATION OF MEMBRANE MATERIAL PROPERTIES

In order to conduct analytical simulations of the problems being studied, it was necessary to have well quantified elastic properties of the Mylar material that was being used for the membrane. This was not a simple matter because of the limited metrology systems available for thin films at that time. For example, the charged-coupled device (CCD) was not invented until 1969. For the membrane experiments conducted in the early 60's at Langley, use was made of non-contacting optical systems. Strain measurements were made using optical micrometers. Two marks were made on the membrane a known distance apart and the displacements of the marks were manually tracked using the optical micrometers. This approach enabled determination of strain. Rotations were measured using small parabolic mirrors ($\sim \frac{1}{4}$ inch square) with a known focal length and a zirconium arc lamp. This type of concentrated arc lamp was used in the U.S. during that time period as a light source intended to deliver an intensely bright point source of white light. By pointing the lamp at the mirror at a distance equal to its focal length and recording the displacement of the reflected light source at that same distance away, the resulting angle of rotation of the mirror could be calculated. This distance of about 20 feet resulted in an effective light lever arm of 40 feet, so that a fairly accurate determination of the angular rotation could be made. Of course currently a small laser could be substituted for the clumsy but effective mirror/lamp system. While the optical micrometers were used to determine strain in axially loaded specimens to determine material modulus and Poisson's ratio, the mirror/lamp system was used to measure rotations in the hub and of the cylinder end bulkheads in the bending tests.

A major source of error in this experimental approach for validating the wrinkled membrane theory was that the Mylar material exhibited an erratic orthotropic behavior that was difficult to quantify with a high degree of accuracy. Extensive quantification of the material's orthotropy was not pursued because the analytical solutions were for isotropic materials only. Thus, best judgment was used in the application of the resulting experimentally determined elastic properties.

Although measurements from both experimental set-ups provided a high degree of confidence in the wrinkled membrane theory, experimental results consistently exhibited a slightly stiffer behavior than predicted by theory. After considerable study, it was believed that the stiffer behavior was a result of the fact that the Mylar material does indeed possess a finite bending stiffness while in the theory the material is assumed to be a perfect membrane with zero bending stiffness. This premise was somewhat corroborated in the work of Reference 2 where the theory was re-derived considering a non-zero stress in the wrinkled region. This new theory had considerably better correlation with test results than the original theory in which the stress in the wrinkled region was zero.

BEHAVIOR OF DOUBLE CURVED PARTLY WRINKLED MEMBRANE STRUCTURES FORMED FROM AN INITIALLY FLAT MEMBRANE

A third wrinkled membrane problem was subsequently investigated in support of very large high altitude helium-filled balloons as well as parachutes. This problem was the behavior of double curved, partly wrinkled membrane structures formed from an initially flat membrane. In both cases, material failures commonly occurred in the crown area of pressurized structures constructed in such a manner. Since the results of linear membrane theory show a stress singularity in the crown area, it is necessary to use nonlinear membrane theory to accurately represent that area. In Reference 6 a nonlinear membrane theory was used in the crown area while a zero hoop stress linear membrane theory was used in the wrinkled region. In Reference 6, the test article and experiment are described in detail. Metrology in this case was an LVDT to measure the out of plane displacement of the membrane under pressure and a simple ruler to measure the extent of the wrinkled region as a function of pressure. Both measurements correlated very well with theory. As with the previous experiments, it is believed that the bending stiffness of the Mylar thin film contributed to relatively small differences between the experimental and theoretical results. Specifically, the extent of the wrinkled region in the physical model was less than that predicted by theory. It is believed that the material bending stiffness, although ever so small, did indeed somewhat mask the apparent full extent of the wrinkled region.

REFERENCES

[1] Stein, M., and Hedgepeth, J. M. "Analysis of Partly Wrinkled Membranes," *NASA TN D-813*, 1961.

[2] McComb, H. G., Jr., Zender, G. W., and Mikulas, M. M., Jr., "The Membrane Approach to the Behavior of Circular Cylindrical Shells in Bending," In *Collected Papers on the Instability of Shell Structures*, NASA TN D-1510, 1962.

[3] Mikulas, M. M., Jr., "Behavior of a Flat Stretched Membrane Wrinkled by the Rotation of an Attached Hub," *NASA TN D-2456*, 1964.

[4] Reissner, E., "On Tension Field Theory," Proc. Fifth Int. Cong. Appl. Mech., John Wiley & Sons, Inc., 1938, pp. 88–92.

[5] Comer, R. L., and Levy, S., "Deflections of an Inflated Circular Cylindrical Cantilever Beam," *AIAA Journal*, Vol. 1, No. 7, July 1963.

[6] Mikulas, M. M., Jr., "Behavior of Double Curved Partly Wrinkled Membrane Structures Formed From an Initially Flat Membrane," Ph.D. Dissertation VPI, June 1970.

INDEX

Note: Page numbers followed by *f* or **t** (indicating figures and tables).

SUPPORTING MATERIALS

A complete listing of titles in the Progress in Astronautics and Aeronautics series is available from AIAA's electronic library, Aerospace Research Central (ARC) at arc.aiaa.org. Visit ARC frequently to stay abreast of product changes, corrections, special offers, and new publications.

AIAA is committed to devoting resources to the education of both practicing and future aerospace professionals. In 1996, the AIAA Foundation was founded. Its programs enhance scientific literacy and advance the arts and sciences of aerospace. For more information, please visit www.aiaafoundation.org.